Recent Techniques and Applications in Ionizing Radiation Research

Edited by Ahmed M. Maghraby and Basim Almayyahi

Published in London, United Kingdom

IntechOpen

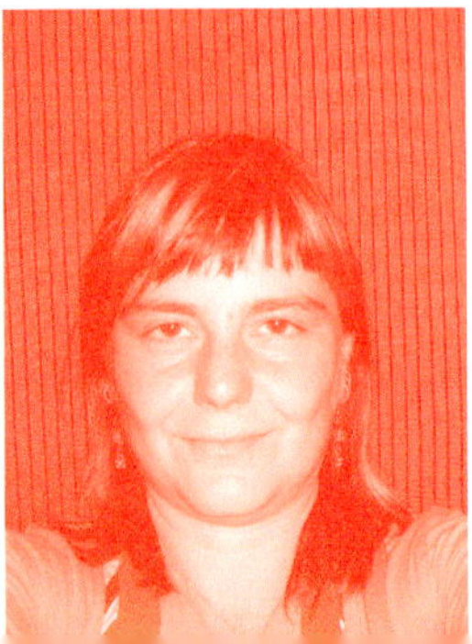

Supporting open minds since 2005

Recent Techniques and Applications in Ionizing Radiation Research
http://dx.doi.org/10.5772/intechopen.87906
Edited by Ahmed M. Maghraby and Basim Almayyahi

Contributors
Entesar H. Elaraby, Nikolay Dolchinkov Todorov, Eugenia Malchukova, Terman Frometa-Castillo, Pooneh Saidi, Mahdi Sadeghi, Adriana Beatriz Beatriz Martínez, Carola Bettina Bettina Bozal, Deborah Ruth Tasat, Nadia Soledad Orona, Angela Matilde Ubios, Ali Farhan Nader Alrekabi, Doaa H. Shabaan, Tayseer I. Al-Naggar, Rajesha Nairy, Nagesh N Bhat, K B Anjaria, Usha Yadav, Rajesh Chaurasia, Kapil Shirsath, Utkarsha Desai, S.K Gupta, B K Sapra, Narayana Yerol, Anil Pyakuryal, Amadeo Wals-Zurita, Asghar Mesbahi

First published in London, United Kingdom, 2021 by IntechOpen
IntechOpen is the global imprint of INTECHOPEN LIMITED, registered in England and Wales, registration number: 11086078, 5 Princes Gate Court, London, SW7 2QJ, United Kingdom
Printed in Croatia

British Library Cataloguing-in-Publication Data
A catalogue record for this book is available from the British Library

Additional hard and PDF copies can be obtained from orders@intechopen.com

Recent Techniques and Applications in Ionizing Radiation Research
Edited by Ahmed M. Maghraby and Basim Almayyahi
p. cm.
Print ISBN 978-1-83962-884-9
Online ISBN 978-1-83962-885-6
eBook (PDF) ISBN 978-1-83962-886-3

For EU product safety concerns:
IN TECH d.o.o., Prolaz Marije Krucifikse Kozulić 3, 51000 Rijeka, Croatia, info@intechopen.com or visit our website at intechopen.com.

Meet the editors

Ahmed M. Maghraby graduated from Cairo University, Faculty of Science, Egypt, in 1995, and started his career as a scientist at the National Institute of Standards (NIS), Egypt. He received his MSc in 1999 and his Ph.D. from Cairo University in 2003 for his study on the calibration and characterization of organic material for use in radiation dosimetry using electron paramagnetic resonance (EPR). In 1999, Dr. Maghraby worked as a guest researcher at the National Institute of Standards and Technology (NIST) in Maryland, USA, and as a visiting scientist at Dartmouth College (EPR center of viable system) in New Hampshire, USA, in 2008. In 2006, he worked as a guest researcher for the Physikalisch-Technische Bundesanstalt (PTB), Germany's national metrology institute. He has led and participated in several international research projects on radiation measurements. His research interest is ionizing radiation metrology, especially ionization chambers, EPR, and thermoluminescence dosimetry (TLD). He is a reviewer for several peer-reviewed journals and has supervised a number of master's and Ph.D. theses, and was elected as the vice president of the National Committee of Physics (NPC) in Egypt.

Dr. Basim Almayahi is currently a professor in the College of Science, University of Kufa, Iraq. His fields of interest include radiation biophysics; radiation medical physics; the natural radioactivity of the environment; treatment of prostate cancer using gold nanoparticles; synthesis and applications of silver nanoparticles on bacterial pathogens; biosensors for detecting radon and lead; the natural radioactivity of gamma rays, radon, and alpha particles in tissue, blood, teeth, bones, soil, water, air, food, fertilizer, and building materials; radiation detection and measurements (scintillation detector (Na I (Tl)); high-purity Germanium (HPGe) detector, Exploranium GR-135; 1029 Continuous Radon Monitor; RAD 7 System; SSNTDs CR-39; radiation protection; statistical physics; and environmental statistics (GenStat, SPSS, OrginLab, Sigmaplot). He was a lecturer in the Department of Physics at Babylon University, Iraq, from 2001 to 2004, before taking up a post as a teacher in the Atomic and Nuclear Lab at the University of Kufa from 2005 to 2008. He is Head of the Department of Environment and Director of Scientific Affairs and Cultural Relations at the College of Science, University of Kufa. He has mentored four master's theses and one Ph.D. dissertation. He is the editor and author of three books and a reviewer for more than 200 journals. He is an editorial board member of twenty-three boards, and a member of various institutions. He has published ninety-five scientific papers in domestic and international journals and participated in forty-three scientific congresses and conferences.

Contents

Preface XI

Section 1
Radiation Protection 1

Chapter 1 3
Relative Biological Effectiveness Studies Using 3 MeV Proton Beam from Folded Tandem Ion Accelerator: An Experimental and Theoretical Approach
by Rajesha K. Nairy, Nagesh N. Bhat, K.B. Anjaria, Usha Yadav, Rajesh Chaurasia, Kapil Shirsath, Utkarsha Desai, S.K. Gupta, B.K. Sapra and Narayana Yerol

Chapter 2 17
Biologically Effective Dose (BED) or Radiation Biological Effect (RBEf)?
by Terman Frometa-Castillo, Anil Pyakuryal, Amadeo Wals-Zurita and Asghar Mesbahi

Chapter 3 27
State of Radiation Protection in Bulgaria
by Dolchinkov Nikolay Todorov

Section 2
Natural Radioactivity 39

Chapter 4 41
Natural Radioactive Decay
by Entesar H. Elaraby

Chapter 5 57
Radon in Foods
by Tayseer I. Al-Naggar and Doaa H. Shabaan

Chapter 6 71
Mathematical Expressions of Radon Measurements
by Ali Farhan Nader Alrekabi

Section 3
Applications in Ionizing Radiation 99

Chapter 7 101
Production of the ^{103}Pd via Cyclotron and Preparation of the Brachytherapy Seed
by Pooneh Saidi and Mahdi Sadeghi

Chapter 8 115
Bisphosphonates as Chelating Agents in Uranium Poisoning
by Adriana Beatriz Martínez, Carola Bettina Bozal, Nadia Soledad Orona, Deborah Ruth Tasat and Angela Matilde Ubios

Chapter 9 133
Influence of the Doping Ion Nature and Content on Defect Creation Processes under the Effect of Ionizing Radiation in Aluminoborosilicate Glasses
by Eugenia Malchukova

Preface

Ionizing radiation is the energy released by atoms either in the form of electromagnetic waves (gamma rays and X-rays) or in the form of particles (alpha, beta, and neutrons). In 1895, Wilhelm Roentgen discovered X-rays and one year later in 1896, Henry Becquerel discovered natural radioactivity. Marie Sklodowska Curi chose radioactivity as the subject of her doctoral thesis. She worked with uranium and thorium and cooperated with her husband Pierre in the successful discovery of two new radioactive elements. Consequently, many scientists paved the way for radiation research.

Ionizing radiation research continued extensively to define that new form of energy, to discover the biological effects of radiation, and to find out methodologies for radiation detection and measurements. New applications for ionizing radiation have emerged including cancer treatment and several diagnostic techniques. In industry, many applications involving radioactive isotopes are now used daily.

It seems that there is always a need for collecting new ideas about a certain subject, especially when it is active, growing, and spreading; nowadays, this is the case with radiation research. This book presents several new subjects about and related to ionizing radiation. Nine authors with nine different topics composed this book, which we hope will add value in the long way of ionizing radiation research.

As editors of this book, we are very thankful to the authors and team members who worked with us for their efforts and patience during the preparation of this text. We hope this book will be useful for undergraduate and postgraduate students, researchers, teachers, practitioners, and every individual who has concerns for a healthy and better life.

Ahmed M. Maghraby
National Institute of Standards,
Egypt

Basim Almayyahi
University of Kufa Email,
Iraq

Section 1

Radiation Protection

Chapter 1

Relative Biological Effectiveness Studies Using 3 MeV Proton Beam from Folded Tandem Ion Accelerator: An Experimental and Theoretical Approach

Rajesha K. Nairy, Nagesh N. Bhat, K.B. Anjaria, Usha Yadav, Rajesh Chaurasia, Kapil Shirsath, Utkarsha Desai, S.K. Gupta, B.K. Sapra and Narayana Yerol

Abstract

Proton being the easiest light ion to accelerate and achieve desired beam profile, has been pursued as a popular particulate radiation for therapy applications. In the present study, *Saccharomyces cerevisiae* D7 strain was used to estimate the RBE values of the 3 MeV proton beam, and an attempt was made to derive mathematical formula for calculating RBE value with respect to the dose. Dosimetry studies were carried out using Fricke dosimetry and Semiconductor Surface Barrier detector to calibrate the absorbed doses of Gamma chamber-1200 and Folded Tandem Ion Accelerator respectively. Gold standard cell survival assay and gene conversion assay were used to compare gamma and proton radiation induced cell death and genetic endpoint. Multi target single hit model was used to derive mathematical formula for RBE estimation. The results show a linear survival-dose response after proton radiation and sigmoid survival-dose response after gamma radiation treatment. The calculated RBE value from the survival and gene conversion studies was 1.60 and 3.93, respectively. The derived mathematical formula is very useful in calculating RBE value, which varies from 3.61 to 1.80 with increasing dose. The estimated RBE value from the mathematical formula is comparable with the experimental values. With the help of the present mathematical formulation, RBE value at any dose can be calculated in the exponential and sigmoidal regions of the survival curve without actually extending the experiment in that dose region, which is not possible using conventional methods.

Keywords: *Saccharomyces cerevisiae*, relative biological effectiveness, radiation dose, cell survival and gene conversion

1. Introduction

Biological effects of heavy charged particles on humans play an important role in two different scientific fields; in radiation therapy using protons and heavier ions and in space research for understanding effects on space travelers from galactic cosmic radiation [1]. In addition, the low energy heavy ion accelerators have an important role in basic and applied sciences [2]. Proton being the easiest light ion to accelerate and achieve desired beam profile, has been pursued as a popular particulate radiation for therapeutic applications. Nonetheless, very less has been understood about biological effectiveness of these charged particles. Proton beams can provide highly localized, uniform doses of radiation to tumors, while sparing the surrounding normal tissues, compared with conventional modalities using photons or electrons [3]. In addition to therapeutic applications, energetic proton also finds its presence in space research, neutron dosimetry wherein due to elastic scattering of energetic neutrons lead to (n, p) reaction and creation of low energy protons in the tissues.

The radiobiological studies conforms that equal physical doses of different types of radiation do not produce equal biological effects, because of differences in their energy deposition patterns. This is taken into account by the concept of Relative Biological Effectiveness (RBE). RBE compares the severity of damage induced by a radiation under test, at a dose D_T relative to the reference radiation dose D_R for producing same biological effect. The reference radiation is commonly ^{60}Co-gamma radiation. Generally, the RBE depends on many factors such as the radiation dose, linear energy transfer (LET) at a given tissue depth, dose rate, energy of the radiation, test system and studied biological endpoint. The RBE values of the radiation are very useful in risk estimation during accidental exposure of ionizing radiation (IR) [4]. Revisions in weighting factors for intermediate and very high energy neutrons as well as accelerated protons in the recent ICRP recommendation has drawn more attention to mechanistic approach of studies using radiobiological endpoints.

In the present study, *Saccharomyces cerevisiae* D7 strain was used to study biological effects of 3 MeV proton radiation using cell survival and gene conversion endpoints. The results obtained were compared with standard ^{60}Co gamma radiation. An attempt has been made to estimate RBE value for 3 MeV proton radiation and variation of RBE value as a function of dose with experimental and theoretical formulations. The model organism considered in the study is Saccharomcyces cerevisiae (budding yeast), which is a useful model for higher eukaryotic organisms study and plays a vital role in modern day research. The conservation of many processes such as replication, DNA damage, replication checkpoints and cell cycle control is observed in Saccharomcyces cerevisiae [5]. Additionally, it has been shown that 42% of yeast genes that cause chromosome instability are conserved in humans, demonstrating the importance of yeast in the study of genomic instability and cancer [6]. The prevalence of budding yeast in research today can also be attributed to the low cost at which, experimental procedures can be completed, coupled with its relatively quick doubling time [7].

2. Materials and methods

Gamma source and Dosimetry: ^{60}Co-gamma chamber-1200 supplied by the Isotope Division, Bhabha Atomic Research Center was used for irradiating the cell samples. Fricke dosimetry system was used to calibrate the gamma chamber; the details were given elsewhere [8–15].

Proton Beam Source and Dosimetry: Proton beams are accelerated using the Folded Tandem Ion Accelerator (FOTIA), a facility at Bhabha Atomic Research Centre (BARC), is an electrostatic accelerator with a maximum terminal voltage of 6 MV [2]. Dosimetry of FOTIA was carried out using 2 Silicon Surface Barrier (SSB) detectors, one mounted inside scattering chamber (Monitor detector) and other on sample position (Sample detector). Position of the beam was visualized using a quartz crystal mounted on a movable ladder in a general purpose scattering chamber maintained in ultra-high vacuum. The pencil beam was made to channel through blank position on the ladder and passed through drift tube of length of about 3 meters. The primary beam was then made to pass through titanium window at the end of drift tube. The beam position was again visualized by keeping a quartz plate after the window. The position of beam was adjusted to the center of the window using X-Y steerer magnets and focused using a quadrupole magnet. Once the beam was tuned to desired geometry and position, the ladder in the scattering chamber was moved so that beam passed through a gold foil of 500 ng/cm^2 thickness. The diffused beam facilitated uniform beam profile at the titanium window. An SSB detector inside the scattering chamber kept at an angle from gold foil helped to monitor the fluence. Another SSB detector was positioned at the position where samples could be mounted, simulating the geometry of sample. The detector was provided with a calibrated collimator to reduce count rate and the fluence measurement was normalized between the two detectors. The profile of the beam was measured by scanning the entire area of titanium window. Intensity of beam was adjusted by varying ion source current. LET measurements were done using TRIM software. The position of the monitor detector was adjusted in such a way that the count rate and dead time of the detector are acceptable. Initial signals from the detector were amplified and digitalized using Multi Channel Analyzer (MCA). The number of scattered proton particles (Monitor proton counts) and diffused proton particles (Sample proton counts) were counted for 100 sec with multiple trials to get the ratio. The fluence of the proton beam at source detector was calculated to measure absorbed dose.

Absorbed dose was calculated using the relation [Kraft et al. 1989]

$$\text{Dose} = \left[\left(1.6 \times 10^{-8}\right) \times LET\left(eV\mathring{A}^{-1}\right) \times \phi(\text{particles cm}^{-2})\right] \tag{1}$$

Where fluence represents particles delivered per unit area and LET represents energy transferred per unit length. The LET of the present setup was estimated to be 13 KeV/µm. The fluence of the source detector was measured using

$$\text{Fluence (F) Source Detector} = \frac{[\text{No.of particles on source detector}]}{\left(\pi \times r_S{}^2\right)} \tag{2}$$

Where 'r_s' represents, sample detector collimator radius. Number of particles on sample detector can be calculated by taking ratio between monitor detector and source detector counts

$$\text{ratio} = \frac{N_M}{N_S} \Leftrightarrow N_S = \frac{N_M}{\text{ratio}} \tag{3}$$

Substituting Eqs. (3) and (2) in (1) gives

$$\text{Dose} = \left[\frac{\left(1.6 \times 10^{-8}\right) \times 13 \times N_M}{\left(\pi \times r_S{}^2 \times \text{ratio}\right)}\right] \tag{4}$$

Rearranging Eq. (4), gives

$$\text{N}_M = \left[\frac{Dose \times \pi \times r_S{}^2 \times \text{ratio}}{\left(1.6 \times 10^{-8}\right) \times 13}\right] \tag{5}$$

Eq. (5) was used to calculate required number of monitor detector counts for desired absorbed dose.

Sample preparation and irradiation: A mutant type diploid yeast strain, *Saccharomyces cerevisiae* D7 was used for the study. The genotype of the strain is

$$\frac{a\ ade2-40}{\propto ade2-119}, \frac{trp5-12}{trp5-27}, \frac{ilv1-92}{ilv1-92}$$

The single cell stationary-phase cultures were obtained by growing the cells Yeast extract: Peptone: Dextrose (YEPD) (1%:2%:2%) medium for several generations to a density of approximately 3×10^8 cells mL^{-1}. Cells were washed thrice by centrifugation (2000 g for 5 min) and re-suspended to a cell concentration of 1×10^8 cells mL^{-1} (by counting in heamocytometer) in sterile double distilled water. For proton radiation, the cell suspension was mixed well and exactly 1×10^8 cells were filtered using millipore filter assembly in aseptic condition. The filter paper having cells on the surface was placed inside sterile 3 cm diameter petri dish and irradiated for different doses. For gamma ray irradiation, polypropylene vials were used containing 1×10^8 cells per ml. Cell suspensions were maintained at 0–4° C before and after irradiation till plating.

Survival assay and Gene conversion assay: Treated and untreated samples were suitably diluted and plated in quadruplicate on YEPD agar medium. Plates were incubated for 2–3 days at 30°C in the dark, and the colonies were counted. The gene conversion assay was conducted by plating 1×10^6 cells per plate on Trp^-agar medium and incubated for 72 h at 30°C in the dark, and the colonies were counted.

3. Results and discussion

Calibration of ^{60}Co-1200 Source: In the present study, vials containing Fricke dosimetry solution were exposed to gamma rays for different time interval. Optical absorbance measurements of the dosimeter were done at 304 nm wavelength using a UV–Visible spectrophotometer. The absorbed dose was calculated using optical absorbance and is presented in **Figure 1**. Dose response was considered to be best fit with linear model, with a regression coefficient equal to 0.99. The dose rate of the gamma chamber was determined by the same method and was found to be 51 Gy min^{-1}. The dose calibrations are traceable to the National standards.

Calibration of FOTIA: The beam profile of the FOTIA was measured using a collimated SSB detector. The dosimetry methods followed is presented in materials and methods section, uniformity of the beam at sample position was measured by placing sample detector vertically and horizontally at different positions from the central axis.

Cell Inactivation Studies: The survival response of *Saccharomyces cerevisiae* D7 strain after irradiation with proton and gamma radiation is presented in **Figure 2**. It is clear from **Figure 3** that dose response with proton beam is linear, whereas with gamma radiation it is sigmoid. The sigmoid dose response is due to the multi-track hit processes combined with dose rate dependent molecular repair processes [9–11]. The linear dose–response is due to the lethal damage which leads to cell death even at lower doses. The absence of shoulder indicates absence of sub lethal damage repair in the case of proton radiation, whereas for gamma radiation, the shoulder indicates that most of the induced sublethal damages were easily repaired at lower doses.

The obtained experimental data were fit to multi-target single hit theory and the survival response of gamma and proton radiation were represented as

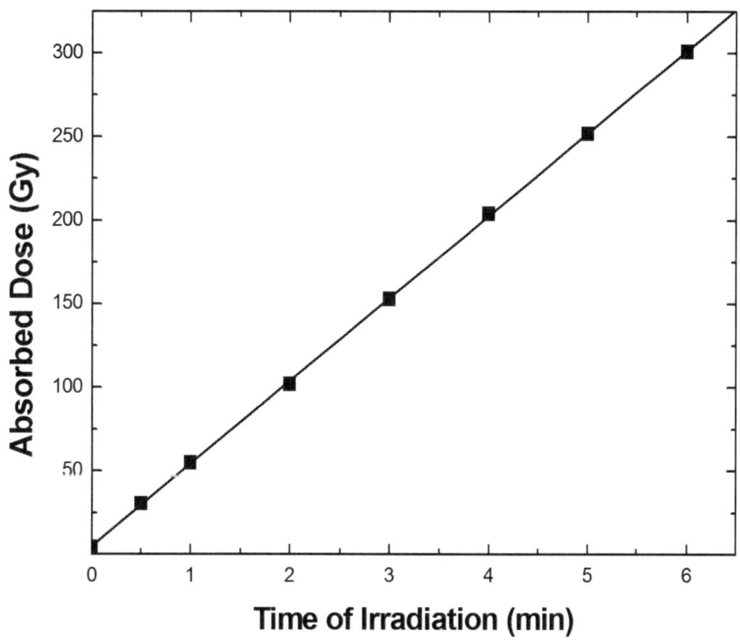

Figure 1.
Dose calibration curve for ^{60}Co gamma Chamber-1200.

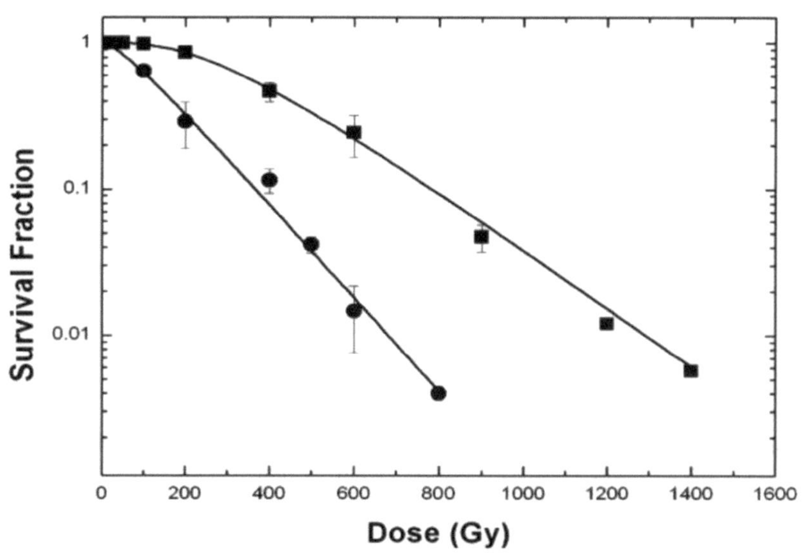

Figure 2.
Dose–response relation after irradiation with proton (●) and gamma radiation (■).

S_{Gamma} = (1-(1-exp(0.00459 D))$^{3.78}$) (with R^2 = 0.99 and Chi^2 = 0.00012) and S_{Proton} = (1-(1-exp(0.00736 D))$^{1.50}$) (with R^2 = 0.99 and Chi^2 = 0.00056). The calculated D_0 value, which is a reciprocal of the inactivation constant, is 218 and 136 Gy for gamma and proton radiation respectively. The RBE value in the exponential region can be calculated by taking ratio between inactivation constant of gamma and proton radiation and is found to be 1.60.

Gene conversation studies: Gene conversion analysis was carried out using *Saccharomyces cerevisiae* D7 yeast cell line at trp locus. Each colony represents a gene

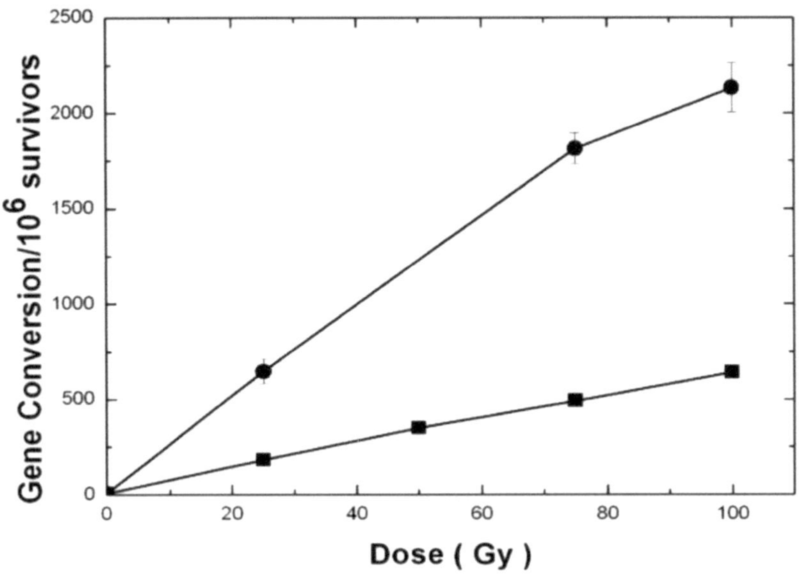

Figure 3.
Gene conversion frequency after irradiating with proton (●) and gamma radiation (■).

convertant and data is presented in **Figure 3**. The doses 25 Gy, 75Gy and 100 Gy were selected and the results show a linear increase in gene conversion frequency with dose. In the case of proton radiation, at lower doses increase in gene conversion frequency was linear, whereas at higher doses it attains plateau. In the case of gamma radiation gene conversion frequency was linear throughout the selected dose region. The gene conversion frequency (G.C.F) for gamma and proton radiation were represented as G.C. $F._{gamma}$ = (6.46 ± 2.19) + (6.46 ± 0.105) D (with R^2 = 0.99) and G.C. $F._{Proton}$ = (7.02 ± 3.44) + (25.44 ± 0.520) D (with R^2 = 0.99). The RBE value of the proton radiation for gene conversion was calculated using slopes, is 3.93.

Relative biological effectiveness studies: In the present study, along with cell inactivation and gene conversion studies, we conducted RBE studies for 3 MeV proton radiations. To estimate RBE value, the experiments were repeated using standard gamma radiation (**Figures 2** and **3**). Estimation of RBE value for proton beam is very important in medical treatment planning, where the RBE values should be known with at least 5–10% accuracy. Generally, a standard RBE value 1.1 is applied to the treatment plan. Recently many authors estimated RBE value for proton beam and they observed that there is a drastic change in RBE value near to Brag's-peak [16–30]. High energy protons have an RBE value of about 1.1, however, for low energy protons still sufficient data is not available to conclude the RBE value. In the present study, we used 3 MeV proton radiation, generally using such energy protons one can observe inside tumor during radiotherapy, so present contributions can be used to strengthen the literature data and can be used to improve proton radiotherapy.

Presently RBE values are calculated on the basis of D_0 doses, which give RBE value in the exponential region. In the present study, we formulated an equation, which can be used to calculate RBE value throughout the selected dose-region. Generally RBE is represented by taking the ratio between gamma radiation and test radiation doses, required to produce the same biological effectiveness.

$$\text{RBE} = \frac{D_G}{D_T} \text{At same biological effectiveness} \tag{6}$$

Where, D_G is gamma radiation dose and D_T is test radiation (in this case proton radiation) dose. From multi-target single hit model, the survival can be represented as

$$S = \{1 - [1 - \exp(-kD)]^n\} \tag{7}$$

Where S represents survival fraction, k is inactivation constant, D is dose and n gives number of targets. To calculate RBE value, we are considering same survival level with both the radiations, thus using Eq. (7), we can write

$$S_G = S_T$$
$$\{1 - [1 - \exp(-k_G D_G)]^{n_G}\} = \{1 - [1 - \exp(-k_T D_T)]^{n_T}\} \tag{8}$$

Simplifying (8), considering high radiation dose (D)

$$\{1 - n_G \exp(-k_G D_G)\} = \{1 - n_T \exp(-k_T D_T)\}$$
$$\frac{n_G}{n_T} = \left\{\frac{\exp(-k_T D_T)}{\exp(-k_G D_G)}\right\}$$
$$\Rightarrow \frac{n_G}{n_T} = \exp(-k_T D_T + k_G D_G)$$
$$\ln\left(\frac{n_G}{n_T}\right) = (-k_T D_T + k_G D_G)$$
$$\frac{1}{D_T} \times \ln\left(\frac{n_G}{n_T}\right) = \left\{-k_T + \frac{k_G D_G}{D_T}\right\}$$
$$\left\{\left(\frac{1}{D_T} \ln\left(\frac{n_G}{n_T}\right) + k_T\right) = \frac{(k_G \times D_G)}{D_T}\right\}$$
$$RBE = \frac{D_G}{D_T} = \left\{\left[\frac{1}{(D_T \times k_G)} \times \ln\left(\frac{n_G}{n_T}\right)\right] + \frac{k_T}{k_G}\right\} \tag{9}$$

Eq. (9) gives the relation between RBE and dose. In Eq. (9), the D_T, n_T, k_T represents dose, number of target and inactivation constant under test radiation condition respectively and n_G, k_G represents number of target, inactivation constant under gamma radiation condition respectively. The variance in the measurements was calculated using following equations, in Eq. (9) the k_G, n_G, k_T and n_T are variables

$$(\sigma y)^2 = \left\{\left[\left(\frac{\partial y}{\partial k_G}\right)^2 \times (\sigma_{k_G})^2\right] + \left[\left(\frac{\partial y}{\partial k_T}\right)^2 \times (\sigma_{k_T})^2\right] + \left[\left(\frac{\partial y}{\partial n_G}\right)^2 \times (\sigma_{n_G})^2\right] + \left[\left(\frac{\partial y}{\partial n_T}\right)^2 \times (\sigma_{n_T})^2\right]\right\} \tag{10}$$

where y represents RBE value

$$\left(\frac{\partial y}{\partial k_G}\right)^2 = \left\{\left[\frac{-1}{(k_G{}^2 \times D_T)} \times \ln\left(\frac{n_G}{n_T}\right)\right] - \frac{k_T}{k_G{}^2}\right\}^2 \tag{11}$$

$$\left(\frac{\partial y}{\partial k_T}\right)^2 = \left(\frac{1}{k_G}\right)^2 \tag{12}$$

$$\left(\frac{\partial y}{\partial n_G}\right)^2 = \left(\frac{1}{n_G \times k_G \times D_T}\right)^2 \tag{13}$$

$$\left(\frac{\partial y}{\partial n_T}\right)^2 = \left(\frac{1}{n_T \times D_T \times k_G}\right)^2 \tag{14}$$

Accordingly standard deviation was calculated. **Figure 4** represents RBE value of 3 MeV proton beam at different doses, calculated using Eq. (9). The experimentally calculated RBE value and theoretically calculated RBE values were compared and presented in **Figure 4**. Very good correlation between experimental and theoretical data was observed.

Higher RBE values were observed at lower doses whereas remain constant at higher doses. RBE values varied in a range from 3.61 to 1.80; the maximum value at lower doses is mainly due to the absence of sub-lethal repair processes. In the case of gamma radiation, at lower doses induced damages are repaired but in the case of proton radiation a small dose also creates lethal damages, hence maximum RBE value was observed. At higher doses the damage due to peroxyl radicals and multi-ionizing events lead to lethal damage in gamma radiation, hence RBE value remains constant. Another reason for higher RBE value is energy deposition pattern of the 3 MeV proton radiation. The LET of the gamma radiation is 0.2–0.3 keV μm^{-1}, whereas 3 MeV proton radiation is 13 keV μm^{-1}. The higher RBE values for low energy protons were reported previously [16–30]. Belli et al. [16] has reported that the RBE depends on LET of the proton radiation. They studied SOBP region proton radiation using V79–753B cells. The RBE value for protons with LET 7.7 keV μm^{-1}, 11 keV μm^{-1}, 20 keV μm^{-1}, 30.5 keV μm^{-1}, 34.6 keV μm^{-1} and 37.8 keV μm^{-1} is 2.22 $\pm$ 0.27, 2.88 $\pm$ 0.37, 3.64 $\pm$ 0.41, 5.59 $\pm$ 0.54, 5.06 $\pm$ 0.51 and 4.50 $\pm$ 0.44, respectively [16]. Similar type observation was made by Folkard et al. [17] and reported an RBE value for protons with mean energies of 1.9, 1.15 and 0.76 MeV, using V79 chinese hamster cells. The RBE values for cell survival at 10% survival level are 1.6, 1.9 and 3.36 for protons with track-average LETs of 17, 24 and 32 keV μm^{-1}, respectively.

In another report Mark Andrew [28] observed an RBE value of 2.6 $\pm$ 0.6 for 94 keV, 3.1 $\pm$ 0.4 for 250 keV, 3.9 $\pm$ 0.8 for 390 keV and 2.4 $\pm$ 0.5 for 1.2 MeV protons using V79 cell line. Belli et al. [18] studied four human cell lines, SCC25, SQ20B

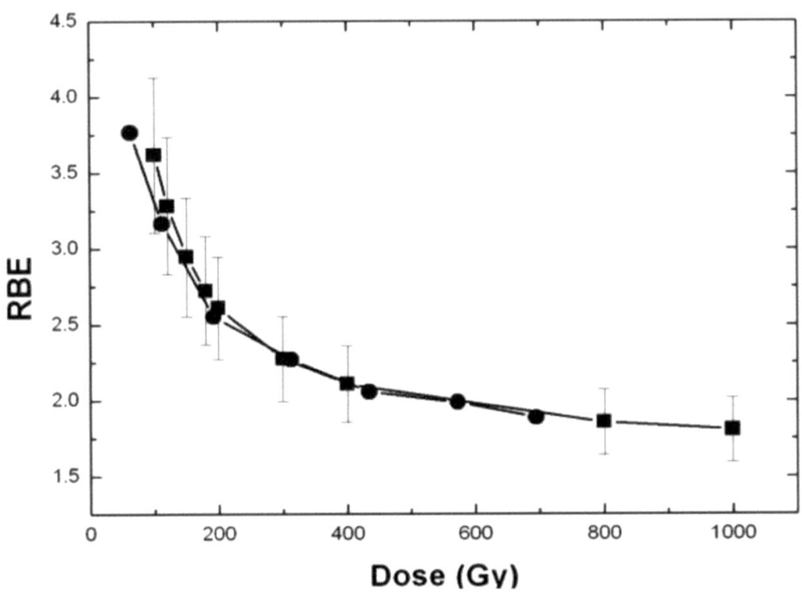

Figure 4.
Variation of RBE with dose; experimental (●) and theoretical (■).

derived from human epithelium tumors of the tongue and larynx, respectively, the normal lines M/10, derived from human mammary epithelium, and HF19 derived from a lung fibroblast. The RBE of the proton beams with LET 30 keV μm^{-1} was 3.2, 1.8, 1.3 and 0.8 for SQ20B, M/10, SCC25, and HF19, respectively [18]. Similarly, Ristić-Fira et al. [29] reported RBE value for mid SOBP region proton particles using radio-resistant human HTB140 melanoma cells and is found to be 2.09 ± 0.36.

Recently, Wéra et al. [30] irradiated Human A549 alveolar adenocarcinoma cells with 4 MeV broad proton beam and calculated RBE value at 10% survival. They reported RBE value of the low energy proton radiation is independent of the dose rate and is equal to 1.9 ± 0.4 for 10 keV μm^{-1} and 2.9 ± 0.5 for 25 keV μm^{-1} [30]. In the same study they calculated RBE values at 77% survival level and were equal to 10.7 ± 3.3 and 3.6 ± 0.6 for 10 keV μm^{-1} and 25 keV μm^{-1} respectively [30]. These values suggest that RBE value depends on survival, which again depends on radiation dose. Britten et al. [22] studied human Hep2 laryngeal cancer cells and V79 cells at various positions along the SOBPs of beams with incident energies of 87 and 200 MeV. Using Hep2 cells, the RBE values were 1.46 at the middle of SOBP, 2.3 at the distal end of the SOBP [22]. For V79 cells, the RBE for the 87 MeV beams was 1.23 for the proximal end of the SOBP, 1.46 for the distal SOBP and 1.78 for the distal end of the SOBP [22]. Similar studies were conducted by Paganetti [23], Słonina et al. [24] and Aoki-Nakano et al. [26] to calculate SOBP region RBE value. They concluded that, the proton RBE value increases with increasing LET which ranges from 1.1 to 4.98. The RBE values for continuous and pulsed proton radiation also studied using human tumor cells [27]. No significant difference was observed between pulsed proton ($RBE = 1.22 \pm 0.19$) and continuous proton ($RBE = 1.10 \pm 0.1$) beam [27].

Previous studies reveal that there is a large variation in reported RBE values among laboratories with the same cell line and a similar LET. For example, Belli et al. [16] and Folkard et al. [17] measured an RBE value of 24 keV μm^{-1} protons as 1.9 and 2.4, respectively. On average, literature reported data concludes RBE value for low energy proton radiation varies from 0.9 to 6, which is comparable with the present findings.

4. Conclusion

The study confirms that, the 3 MeV proton beam is more lethal to biological system compare to gamma radiation and the dose response was found to be linear. Nearly 4 times higher gene conversion frequency was observed in proton radiation as compared to gamma radiation. The estimated RBE value estimated from the mathematical equation developed in the present study is comparable with the experimental values. The RBE value of the 3 MeV protons was found to decreases with the dose and varied from 3.61 to 1.80. With the help of the present mathematical formulation, RBE value at any dose can be calculated in the exponential region of the survival curve without actually extending the experiment in that dose region, which is not possible using conventional methods.

Acknowledgements

The authors from Mangalore University are thankful to Board of Research in Nuclear Sciences, Department of Atomic Energy, Government of India, for the financial support.

Author details

Rajesha K. Nairy[1*], Nagesh N. Bhat[2], K.B. Anjaria[2], Usha Yadav[2], Rajesh Chaurasia[2], Kapil Shirsath[2], Utkarsha Desai[2], S.K. Gupta[3], B.K. Sapra[2] and Narayana Yerol[4]

1 Department of Physics, P.C. Jabin Science College, 580031, Hubballi, Karnataka, India

2 RP and AD, Bhabha Atomic Research Center, 400085, Mumbai, India

3 IADD, Bhabha Atomic Research Centre, 400085, Mumbai, India

4 Department of studies in Physics, Mangalore University, 574 199, Mangalagangotri, India

*Address all correspondence to: rajesh.nairy@gmail.com

References

[1] Oliver J. The relative biological effectiveness of proton and ion beams. Z. Med.Phys. 2008;**18**(4):276-285

[2] Singh P. Folded tandem ion accelerator facility at BARC. Pramana-Journal de Physique. 2001;**57**:639

[3] Kim SS, Choo DW, Shin D, Baek HJ, Kim TH, Motoyama N, et al. In vivo radiobiological characterization of proton beam at the National Cancer Center in Korea: Effect of the Chk2 mutation. International Journal of Radiation Oncology, Biology, Physics. 2011;**79**(2):559-562

[4] Nikjoo H, Lindborg L. RBE of low energy electrons and photons. Physics in Medicine and Biology. 2010;**55**(10):R65-R109

[5] Johnson A, O'Donnell M. Cellular DNA replicases: Components and dynamics at the replication fork. Annual Review of Biochemistry. 2005;**74**: 283-315

[6] Yuen KW, Warren CD, Chen O, Kwok T, Hieter P, Spencer FA. Systematic genome instability screens in yeast and their potential relevance to cancer. Proceedings of the National Academy of Sciences of the United States of America. 2007;**104**(10): 3925-3930

[7] Botstein D, Fink GR. Yeast: An experimental organism for modern biology. Science. 1988;**240**(4858): 1439-1443

[8] Fricke H, Hart E J, In; Radiation Dosimetry, Vol II, Eds. Allix F. H. and S. C. Academy Press, W. C. New York, 1968.

[9] Nairy R, Bhat NN, Anjaria KB, Sreedevi B, Sapra BK, Narayana Y. Study of gamma radiation induced damages and variation of oxygen enhancement ratio with radiation dose using Saccharomyces cerevisiae. Journal of Radioanalytical and Nuclear Chemistry. 2014;**302**:1027-1033

[10] Joseph P, Acharya S, Sanjeev G, Bhat NN, Narayana Y. Cell inactivation studies on yeast cells under euoxic and hypoxic condition using electron beam from microtron accelerator. Journal of Radioanalytical and Nuclear Chemistry. 2011;**290**(1):209-214

[11] Joseph P, Nairy R, Acharya S, Ganesh S, Narayana Y. Chemical dosimeters for electron beam dosimetry of microtron accelerator. Journal of Radioanalytical and Nuclear Chemistry. 2014;**302**:1013-1019

[12] Chaurasiaa KR, Balakrishnana S, Kunwarb A, Yadav U, Bhat N, Anjariaa K, et al. Cyto-genotoxicity assessment of potential radioprotector,3,3-diselenodipropionic acid (DSePA) in Chinese hamster ovary (CHO) cells and human peripheral blood lymphocytes. Mutation Research/ Genetic Toxicology and Environmental Mutagenesis. 2014;**774**:8-16

[13] Joseph P, Narayana Y, Nairy R, Ganesh S, Bhat NN. Assessment of electron and gamma induced DNA damage in human peripheral blood by alkaline comet assay. Radiation Protection and Environment. 2011; **34**(4):221-224

[14] Nairy RK, Bhat NN, Joseph P, Sanjeev G, Yerol N. Studies on electron beam induced DNA damage and repair kinetics in lymphocytes by alkaline comet assay. Int. Journal of Radiation Research. 2015;**13**(3):213-220

[15] Nairy RK, Bhat NN, Joseph P, Sanjeev G, Yerol N. Dose response study using Micronuclueus cytome assay-a tool for biodosimtry application. Radiation Protection Dosimetry. 2016

[16] Belli M, Cera F, Cherubini R, Dalla Vecchia M, Haque AMI, Ianzini F, et al. RBE–LET relationships for cell inactivation and mutation induced by low energy protons in V79 cells: Further results at the LNL facility. International Journal of Radiation Biology. 1998;**74**: 501-509

[17] Folkard M, Prise KM, Vojnovic B, Davies S, Roper MJ, Michael BD. The irradiation of V79 mammalian cells by protons with energies below 2 MeV. Part I: Experimental arrangement and measurements of cell survival. International Journal of Radiation Biology. 1989;**56**(3):221-237

[18] Belli M, Bettega D, Calzolari P, Cera F, Cherubini R, Dalla Vecchia M, et al. Inactivation of human normal and tumour cells irradiated with low energy protons. International Journal of Radiation Biology. 2000;**76**(6):831-839

[19] Bettega D, Calzolari P, Chauvel P, Courdi A, Herault J, Iborra N, et al. Radiobiological studies on the 65 MeV therapeutic proton beam at Nice using human tumour cells. International Journal of Radiation Biology. 2000; **76**(10):1297-1303

[20] Paganetti H, Niemierko A, Ancukiewicz M, Gerweck LE, Goitein M, Loeffler JS, et al. Relative biological effectiveness (RBE) values for proton beam therapy. International Journal of Radiation Oncology, Biology, Physics. 2002;**53**(2):407-421

[21] Butterworth KT, McGarry CK, Clasie B, Carabe-Fernandez A, Schuemann J, Depauw N, et al. Relative biological effectiveness (RBE) and out-of-field cell survival responses to passive scattering and pencil beam scanning proton beam deliveries. Physics in Medicine and Biology. 2012;**57**(20): 6671-6680

[22] Britten RA, Nazaryan V, Davis LK, Klein SB, Nichiporov D, Mendonca MS, et al. Variations in the RBE for cell killing along the depth-dose profile of a modulated proton therapy beam. Radiation Research. 2013;**179**(1):21-28

[23] Paganetti H. Relative biological effectiveness (RBE) values for proton beam therapy. Variations as a function of biological endpoint, dose, and linear energy transfer. Physics in Medicine and Biology. 2014;**59**(22):R419-R472

[24] Słonina D, Biesaga B, Swakoń J, Kabat D, Grzanka L, Ptaszkiewicz M, et al. Relative biological effectiveness of the 60-MeV therapeutic proton beam at the Institute of Nuclear Physics. (IFJ PAN) in Kraków, Poland. Radiation and Environmental Biophysics. 2014;**53**(4): 745-754

[25] Matsumoto Y, Matsuura T, Wada M, Egashira Y, Nishio T, Furusawa Y. Enhanced radiobiological effects at the distal end of a clinical proton beam: in vitro study. Journal of Radiation Research. 2014;**55**(4):816-822

[26] Aoki-Nakano M, Furusawa Y, Uzawa A, Matsumoto Y, Hirayama RT, suruoka C, et al. Relative biological effectiveness of therapeutic proton beams for HSG cells at Japanese proton therapy facilities. Journal of Radiation Research. 2014;**55**(4):812-815

[27] Zlobinskaya O, Siebenwirth C, Greubel C, Hable V, Hertenberger R, Humble N, et al. The effects of ultra-high dose rate proton irradiation on growth delay in the treatment of human tumor xenografts in nude mice. Radiation Research. 2014;**181**(2):177-183

[28] Mark Andrew S. Fast neutron relative biological effectiveness determination via proton bombardment of V79 cells master of science thesis submitted to Massachusetts institute of. Technology. 1996

[29] Ristić-Fira A, Todorović D, Zakula J, Keta O, Cirrone P, Cuttone G, et al.

DOI: http://dx.doi.org/10.5772/intechopen.94243

Response of human HTB140 melanoma cells to conventional radiation and hadrons. Physiological Research. 2011; **60**(1):S129-S135

[30] Wéra AC, Heuskin AC, Riquier H, Michiels C, Lucas S. Low-LET proton irradiation of A549 non-small cell lung adenocarcinoma cells:Dose response and RBE determination. Radiation Research. 2013;**179**(3):273-281

Chapter 2

Biologically Effective Dose (BED) or Radiation Biological Effect (RBEf)?

Terman Frometa-Castillo, Anil Pyakuryal, Amadeo Wals-Zurita and Asghar Mesbahi

Abstract

The current radiosensitive studies are described with linear-quadratic (LQ) cell survival (S) model for one fraction with a dose d. As result of assuming all sublethally damaged cells (SLDCs) are completely repaired during the interfractions, that is, no presence of SLDCs, the survived cells are calculated for a n-fractionated regimen with the LQ S(n,D) model. A mathematically processed subpart of LQS(n,D) is the biologically effective dose (BED) that is used for evaluating a so-called "biological dose." The interactions of ionizing radiation with a living tissue can produce partial death or sublethal damage from healthy or sublethally damaged cells. The proportions of the killed and sub-lethally damaged cells define the radiation biological effects (RBEfs). Computational simulations using RBEFs for fractionated regimens let calculating tumor control probability. While the derivation of the LQ S(n,D) considers a 100% cell repair, that is, 0% of sublethally damaged cells (SLDCs), the radiobiological simulators take into account the presence of SLDCs, as well as a cell repair <100% during the interfractions and interruption. Given "biological dose" does not exist, but RBEf, there was need for creating the BED. It is shown how some uses of BED, like the derivation of EQ2D expression, can be done directly with the LQ S(n,D).

Keywords: BED, simulation, radiotherapy, brachytherapy, fractionation, linear-quadratic model, mathematical models, radiobiology

1. Introduction

In 1989, an article published in [1] introduced the term BED, biologically effective dose, as a linear-quadratic (LQ)-based formula. After 21 years, a new article was published in [2] for showing the wide use of BED in the radiation therapies. In this work, the BED was defined (of a given schedule) as: "the total dose required to give the same log cell kill as the schedule being studied, at an infinitely low dose-rate or with infinitely small fractions well-spaced out; now with an overall time factor for repopulation during continued irradiation."

When ionizing radiation interacts with a determined volume of living tissue, this can or cannot interact with all cells, can or cannot produce effects as result of their interactions; and the first fraction of a fractionated treatment produces a partial number of killed and sublethally damaged cells (SLDCs) from the total initially

 IntechOpen

undamaged ones. During the second and successive fractions, the radiation can interact with these three kinds of cells, where the interactions can produce the same effects of the first fraction from the undamaged cells and SLDCs.

BED has direct relationship with the radiation biological effects (RBEfs), in particular the cell survival (S) in radiation treatments with n fractions, d dose per fractions delivered in tissues characterized with LQ parameters α and α/β. The BED expression was a result of a mathematical derivation of the exponential part of the LQ S model for treatments with n fractions and dose per fraction d, the LQ S(n,D) where D = nd; and this model was obtained assuming that all sublethally damaged cells are wholly repaired during the interfraction period.

"BED is a measure of the true biological dose delivered by a particular combination of dose per fraction and total dose to a particular tissue characterized by a specific α/β ratio" [3]. This expression is incoherent because only the physical dose is delivered, and produces biological effects. There is no "biological dose."

The mathematical formulas have traditionally been used for calculating physical quantifications of deterministic and stochastic processes/effects (SP/Es). At this time there is high development in the computer science, where the computational simulators allow us determining probabilistic metrics some SP/Es, such as their means and probabilities, based on simulations of many possible cases.

The RBEfs should be estimated with computational radiobiological simulators or with the current LQ S(n,D) model.

2. The biologically effective dose (BED)

Nowadays, the radiosensitivity studies function of the absorbed dose (d) are described with the cell survival (S), which is complement of cell kill (K), and probabilistically S = 1-K. These studies are widely modeled with the LQ S(d) for one fraction as

$$LQS(d) = exp\left(-\alpha d - \beta d^2\right) \tag{1}$$

where α and β are the LQ parameters.

d: dose of one fraction.

As result of assuming all sublethally damaged cells (SLDCs) are completely repaired during the interfractions, that is, no presence of SLDCs, the survived cells are calculated for a n-fractionated regimen as

$$LQS(n, D) = [LQ\ S(d)]^n = exp\left(-\alpha D - \beta D^2/n\right) \tag{2}$$

where D = n*d.

A mathematically processed subpart of LQ S(n,D) is the BED that is used for evaluating a called "biological dose," and is written as

$$BED = D\left[1 + \frac{d}{\alpha/\beta}\right] \tag{3}$$

As an inherent part of the LQ S(n,D) model, the origin of BED is explained in [4] the following way.

1. The radiation cell kill (or effect, E) can be expressed as

$$Cell\ kill = E = n\left(\alpha d + \beta d^2\right) \tag{4}$$

2. Consider a progressive reduction in d such that it approaches a value of zero. Although the number of fractions n will then need to be increased to maintain the same effect, βd_2 will be very small in comparison with ad (since d will greatly exceed d_2 for very small values and α always exceeds β). Therefore, when d is very small, Eq. (4) is approximated as

$$E = n\alpha d = \alpha D \tag{5}$$

3. This demonstrates that the total dose (D) of radiotherapy given at a very low dose per fraction represents the highest total dose required to obtain a specific effect. The total dose required in these conditions constitutes the definition of BED in situations where cellular repopulation can be ignored, that is, in this limiting case:

$$BED = D = E/\alpha \tag{6}$$

The authors of [4] considered that: "BED represents the physical dose required for a given effect if the dose were to be delivered by infinitely small doses per fraction or, in the case of continuous radiation rates, at a very low dose rate."

The procedure used in [4] is purely mathematical, since the cell kill (K) and its complement, the cell survival (S), are stochastic effects with a deterministic region for low values of d, where D does not produce any effect, that is, there will be 100% of S; i.e. 0% of K.

To date, except the probabilistic treatments of the tumor control/normal tissue complication probability (TCP/NTCP), many stochastic processes/effects in areas of the ionizing radiations interacting with living tissues have not been probabilistically treated nor modeled, which has led deficiencies, like replacement in the evaluations of cell survival (S)—a probabilistic metric—by BED, a non-probabilistic, a mathematical derivation from the LQ S(n,D) formalism.

2.1 The BED in radiotherapy

In [4], the authors have shown the use of BED in practical situations for normal tissues. For example, if a dose of 60 Gy in 30 fractions is received by a critical normal tissue, the associated BED may (for example) be expressed in terms of Gy1.5, Gy2, and Gy3 (for α/β ratios of 1.5, 2, and 3 Gy). The initial BED value for a fractionation schedule of 60 Gy in 30 fractions (BED = $140Gy_{1.5}$) is used to calculate the total dose and dose per fraction for the alternative schedule of 20 fractions. The result for alternative fractionation schedule is obtained from the solution of d in a rearrangement of following equation

$$20d\left(1 + \frac{d}{1.5}\right) = 140Gy_{1.5} \tag{7}$$

Really, we can use the Eq. (2) for determining the previous alternative fractionation schedule (n2 fractions and d2 dose per fractions) without need of creating the BED, based on the following procedure

$$LQS(n_1, D_1) = LQS(n_2, D_2) \tag{8}$$

where $D_1 = n_1d_1$ and $D_2 = n_2d_2$

$$exp\left(-\alpha D_1 - \beta {D_1}^2/n_1\right) = exp\left(-\alpha D_2 - \beta {D_2}^2/n_1\right) \tag{9}$$

$$-\alpha D_1 - \beta {D_1}^2/n_2 = -\alpha D_2 - \beta {D_2}^2/n_1 \tag{10}$$

On multiplication of Eq. (10) by $-1/\alpha$, then

$$n_1d_1 + n_1\frac{{d_1}^2}{\alpha/\beta} = n_2d_2 + n_2\frac{{d_2}^2}{\alpha/\beta} \tag{11}$$

Substituting $n_1 = 30$, $d_1 = 2$ Gy ($D_1 = n_1d_1 = 60$ Gy), and $n_2 = 20$, we obtain the same Eq. (7) but without the dimension $Gy_{1.5}$

$$20d\left(1 + \frac{d}{1.5}\right) = 140 \tag{12}$$

From the Eq. (11) one can derive the current equivalent dose in 2-Gy fractions (EQD2) in Gy, that is, the Eq. (14), if one substitutes $n_1 = n_2$ and $d_2 = 2Gy$ transform the Eq. (11) as

$$D_1 + D_1\frac{2Gy}{\alpha/\beta} = D_2 + D_2\frac{d}{\alpha/\beta} \tag{13}$$

where $D_1 = EQD2$ and $D_2 = D = nd$, then

$$EQD2 = D(d + \alpha/\beta)/(2Gy + \alpha/\beta) \tag{14}$$

This derivation does not need creation of the BED.

With the introduction of BED in radiotherapy, the radiation biological effects of radiation treatments have been characterized with BED with generic values $\alpha/\beta = 10$ Gy for tumors and $\alpha/\beta = 3$ Gy for normal tissues.

While the BED expression, the Eq. (3), has only one parameter, α/β, the LQS(n,D) has two: α and β. Therefore, a tissue with $\alpha = 1\ Gy^{-1}$ and $\alpha/\beta = 10$ Gy that receives 60 Gy in 30 fractions, that is, d = 2 Gy, will have a biological radiation effect of 9.1% of cell survival.

The cell repopulation (CR) has been introduced in Eq. (3) as

$$BED = D\left[1 + \frac{d}{\alpha/\beta}\right] - K(T - T_K) \tag{15}$$

where T is the overall treatment duration.

T_k is the time when the cell repopulation starts.

K is the factor in Gy/day. According to [4], it is the daily BED equivalent of repopulation.

The CR can be introduced in Eq. (2) as

$$LQS(n, D) = exp\left(-\alpha D - \beta D^2/n\right) + KS(T - T_K) \tag{16}$$

where KS is the factor in 1/day that represents the rate of the CR per day.

The authors of [4] have highlighted that BEDs are additive. It means that if radiotherapy is given in multiple phases, then the BED for each phase can be summated to give the total BED.

Actually, the RBEfs are additive, that is, the biological damages increase when number of irradiation phases increase.

2.1.1 The BED in interrupted treatments

Many of the current works, such as [5–8] related with the interrupted treatment use directly the BED expression or with some modifications that involve elements, like the cell repopulation.

The BED is one of the most current important tools for compensating interrupted radiation treatments, where, as described in [5] three values of BED (original for the initial prescription, applied before the interruption and a new for compensating the interruption) are considered.

2.2 The BED in brachytherapy (BT)

The BT may be delivered at high, medium, or low dose rates, respectively HDR, MDR, or LDR.

The BED is also expressed as the product of the total physical dose (D) and a dimensionless factor RE as

$$BED = D * \text{RE}. \tag{17}$$

Within of the BED expression, they have tried of including other factors affecting the RBEfs, such as cell sublethal damage (SL), cell repair (Rt), and repopulation (Pt). These inclusions are notary in the BT as shown in [9].

The works of [9–11] are strongly based on the BED. Here the factors affecting RBEf are added in the BED expression or included in its dimensionless subpart, the RE. More than 16 equations modifying Eq. (17) have been developed in this work. The Pt effect is considered in the BED expression as:

$$BED = D * RE - RCF \tag{18}$$

$$RCF = K * \left(T - T_{delay}\right) \tag{19}$$

where T is the overall time, T_{delay} is the delay time after the beginning of treatment before the repopulation rate becomes significant, and K is a parameter of this model.

The Rt effect for a continuous low dose rate is considered into RE as a complex expression in [9]. In the same reference, for permanent implant, the decay of the radioactive sources is incorporated in the factor RE of Eq. (17) as

$$RE = 1 + \frac{R0}{(\mu + \lambda)\left(\alpha/_{\beta}\right)} \tag{20}$$

where R0 is the initial dose rate, and λ is the radionuclide decay constant.

3. The radiation biological effect (RBEf)

After the first fraction of irradiation to a living tissue region with a dose d, a number of killed and sublethally damaged cells appear. The mean outcomes of the radiation interactions with the cells are probabilistically related as

$$K + SL + U = 1\,(100\%) \tag{21}$$

$$S = SL + U \tag{22}$$

$$K + S = 1\,(100\%) \tag{23}$$

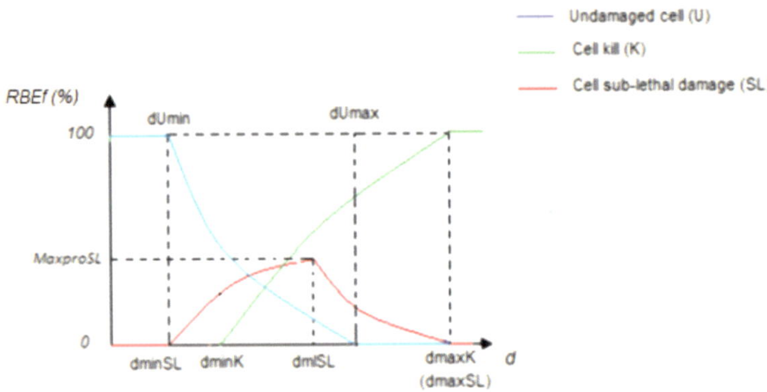

Figure 1.
Representation of the radiation biological effects (RBFs) defined by cell kill (K) and sublethally damaged cell (SL), and undamaged cell. Abbreviations: dUmin and dUmax: Respectively the lower and upper limit for stochastic region of U; dKmin and dKmax: Respectively the lower and upper limit for stochastic region of K; dminSL and dmaxSL: Respectively the lower and upper limit for stochastic region of SL; dmlSL: Value of d with maximum of SL; MaxproSL: Maximum value of SL.

$$K = 1 - S \tag{24}$$

where K is the mean cell kill; SL is the mean cell sublethal damage; U is the mean undamaged cell; and S is the mean cell survival. **Figure 1** is a representation of RBEfs defined by the mean values of the cell kill, sublethally damaged cell, as well undamaged cell. These are the immediate results of first fraction of irradiation with a dose d in a living tissue.

4. The radiobiological computational simulators (RCSs)

The computational simulations have led to the development of three radiobiological simulators that determine TCP and mean RBEFs in normal tissue for regular/interrupted treatments, as well as one that obtains similar probabilistic distributions to binomial and Poisson ones. The first application is discussed in [12]. The MatLab applications of these simulators are publicly available in the repository of [13].

The TCP has been traditionally obtained from experimental/observational data, complex phenomenological/mechanistic models as shown in [14]. The TCP computational-calculation methodology simulates possible situations of an irradiated tumor, and is based on probabilistic analysis of three possible kinds of cells and their final results during the interactions for a tumor homogeneously irradiated in a fractionated regimen. The cell repair is taken into account as a temporal process during the interfractions.

Given there will be tumor control when tumor cells are all killed by radiation, it allows to determine TCP based on its probabilistic definition in the computational simulations.

In the region with the minimum dose per fraction of a tumor heterogeneously irradiated, there is the highest value of the cell survival shown by Eq. (1); that is, there is the lowest value of probability of cell kill. For this reason, the TCP should be calculated analyzing the results of interactions in this region, and it is not necessary to consider other tumor regions.

For simulating a fractionated/interrupted treatment, it is considered the following:

- The first fraction generates a mean *nkc* killed cells, *nslc* sublethally damaged cells, and *nudc* undamaged cells from the total cells *NTC*.
- For the second and successive fractions, the three kinds of cells are analyzed in their possible final outcomes in each fraction.
- The radiation can interact with a killed cell or a survived cell. If a random number *gnum* is generated, and *gnum* < = *nkc/NTC*, then the cell is killed, but is survived.
- For a killed cell, the simulator will analyze a new cell; but for a sublethally damaged cell, there are two possibilities: the cell is undamaged or sublethally damaged. This is defined with a new *gnum* > *nslc/(nslc + nudc)* for a undamaged cell.
- For an undamaged cell, if a new gnum < probability for cell kill (K), this cell will die, but if gnum <= (K+ probability for cell sublethal damage), this is become in a sublethally damaged.
- For a sublethally damaged cell (SLDC) there is a range of damage degree. Two new random numbers gnum1 and gnum2 are generated, and let us define *KSL = max(gnum1;1-gnum1)*. *If gnum2* < =*KSL*, the cell will die, but is kept as a SLDC. The previous condition is associated to a major probability of killing the SLDC.
- While the number of fractions increases, nkc increases, and nudc decreases. The nslc can increase or decrease after the second and successive fractions.
- The number of repaired cells is determined after each fraction or during an interruption.
- TCP is calculated as ratio of simulations with nkc = NTC and total of them.
- The radiosensitivity for cell kill (K) is calculated from Eq. (1) as K = 1 − LQS(d)
- Eq. (22) shows that survived cells involve sublethally damaged and undamaged cells. The current radiosensitivity studies only report mean values of probability for S that is the sum of probabilities for SL and U, so we have assumed the probability for SL as SL < = S in our radiobiological simulators.

Eq. (1) represents cell survival probability, that is, mean value of the ratio of the sublethally damaged cells and total of them, when a determined living tissue characterized with parameters α and α/β is homogenously irradiated with one fraction of dose d. For this reason, this equation can be considered for whatever healthy cell of a determined tissue as probability of becoming in survived cell after irradiation. Given cell kill is a probabilistic complement of cell survival, then cell kill probability is equal to

$$K(q) = 1 - LQ\ S(d) \tag{25}$$

Our computational simulations have not been previously applied by the current Monte Carlo methods. In the radiobiological modeling and simulation field applied to radiotherapy, this methodology will represent a big contribution due to one potential innovation being that rather than evaluating TCP by analytically calculating, the TCP is calculated based on its own probabilistic definition.

Contrary to TCP, the normal tissue complication probability (NTCP) calculation does not have easy way for being determined in the RCSs. Therefore, we have suggested assuming similar-Poisson distributions for the NTCPs, and evaluating safety in the radiation treatments with $NTCP_0$ (normal tissue non-complication probability).

4.1 The RCS in radiotherapy

In [15], authors recognize that BED formula does not take into account altered fractionation, like twice-daily fractionation. The radiobiological simulators do not have the quoted limitation of the BED. Using computational simulations one can simulate any schedule of fractionation.

Example 1. A normal tissue (NT) region that is characterized with $\alpha = 0.0683Gy^{-1}$ and $\alpha/\beta = 1.5Gy$; for $d = 0.1Gy$, its radiosensitivity for cell sublethal damage is equal to 1%. As result of simulating this NT region with cell repair equal to 40% and cell density 10^7cells/cm^3, in 30 fractions, the mean cell kill is 30% and mean cell sublethal damage is 0.249%.

As external beam radiotherapy, BT is an activity that involves interactions of ionizing radiations with living tissues, which produce RBEfs into these tissues. The specific treatment duration will depend on many different factors, including the required rate of dose delivery and the type, size, and location of the cancer, and is still calculated from prescribed dose.

Although the authors have not developed radiobiological simulators for BT, the methodology of these tools can be extended to this radiation therapy.

4.1.1 The RCS in interrupted treatments

An interrupted treatment is a fractionated one, where there is a long-time period greater than the normal interfractions. During the interruption, the sublethally damaged cells have a major possibility of being repaired than during the interfractions of a regular treatment.

Nowadays, the interrupted treatment is evaluated with only a radiobiological tool, the BED. The implementation of the RCSs in the interrupted treatments will represent an extension of the new methodology already applied for the regular treatment.

Cell repopulation, like cell repair, is one of the temporal cellular processes, and is related with the tumor growth, so for an interrupted treatment should be compensating with an increase of the field of the radiation beam.

5. Conclusions

While the derivation of the LQ S(n,D) model for fractionated regimen considers a 100% cell repair, that is, 0% of sublethally damaged cells (SLDCs), radiobiological simulator methodology takes into account the presence of SLDCs, as well as a cell repair <100% during the interfractions. This makes a better real simulation of the process of interaction of ionizing radiation with living tissues.

The BED is a virtual and redundant radiobiological concept because of this being just a processed subpart of the LQ S(n,D) model by means of a mathematical derivation, and its expression does not model neither physical nor biological

$$C_{Ra} = \frac{\rho\, A\, h}{K\, MT_{eff}} \tag{39}$$

Since the unit of time in exhalation formula in hour, we should convert units of λ, T_{eff} and K to hour. All the quantities in **Table 2** are known except (C_I, C, C_{avg}). We can find ρ experimentally by using CR-39 or LR-115 type II based on radon dosimeter and by using continuous radon monitor to determine the value of (C_I, C, C_{avg}). From **Table 2**, we show that the quantities $(\lambda, T, T_{eff}, K, M, A, V)$ are constants, therefore the relationship between radon exhalation rate and effective radium content with the quantities (C_I, C, C_{avg}, ρ) are linear. From the relations in **Table 2**, we get some relations which show the ship linear between these quantities [1]:

$$\left.\begin{aligned} \frac{E_A}{E_M} &= \frac{M}{A} \\ \frac{E_A}{C_{Ra}} &= \lambda \frac{M}{A} \\ \frac{E_M}{C_{Ra}} &= \lambda \end{aligned}\right\} \tag{40}$$

2.5 Closed-can technique (two different detectors)

In this technique using two different SSNTD detectors (CR-39 & LR-II) were separately placed in close-can, to measurement and discriminate between radon and thoron concentrations which escape from sample, at same time **Figure 2**. During this exposure the α-particles emitted by the nuclei of radon and thoron and its progenies have bombarded the SSNTD films [1, 8].

The global track density rates, due to α-particles emitted by radon, registered on the CR-39 detector ρ^G_{Rn} can be writing mathematical:

$$\rho^G_{Rn} = \rho_{Rn} + \rho_{Po218} + \rho_{Po214} \tag{41}$$

where ρ_i are the track density for radon and its progenies on CR-39 detectors and from Eq. (19) can written by concentration average of radon and its progenies:

$$\rho^G_{Rn} = K_{Rn} C_{Rn}\, T + K_{Po218} C_{Po218} T + K_{Po214} C_{Po214} T \tag{42}$$

where K_i are the calibration factors for radon and its progenies. By malty and dividing right hand-Eq. (42) C_{Rn}, we obtain:

$$\rho^G_{Rn} = \left(K_{Rn} + K_{Po218} \frac{C_{Po218}}{C_{Rn}} + K_{Po214} \frac{C_{Po218}}{C_{Rn}} \right) C_{Rn} T \tag{43}$$

Quantity	C_I	C_s	C_{avg}	ρ
$E_A\left(Bq\, m^{-2}\, h^{-1}\right) =$	$\frac{C_I\, \lambda\, V}{A\, T_{eff}}$	$\frac{C_s\, \lambda\, V}{A}$	$\frac{C_{avg} T\, \lambda\, V}{A\, T_{eff}}$	$\frac{\rho\, \lambda\, V}{KA\, T_{eff}}$
$E_M\left(Bq\, kg^{-1}\, h^{-1}\right) =$	$\frac{C_I\, \lambda\, V}{M\, T_{eff}}$	$\frac{C_s\, \lambda\, V}{M}$	$\frac{C_{avg} T\, \lambda\, V}{M\, T_{eff}}$	$\frac{\rho\, \lambda\, V}{KMT_{eff}}$
$C_{Ra}\left(Bq\, kg^{-1}\right) =$	$\frac{C_I V}{MT_{eff}}$	$\frac{C_s\, V}{M}$	$\frac{C_{avg} T\, V}{MT_{eff}}$	$\frac{\rho\, V}{K\, MT_{eff}}$

Table 2.
The exhalation radon rate formulize.

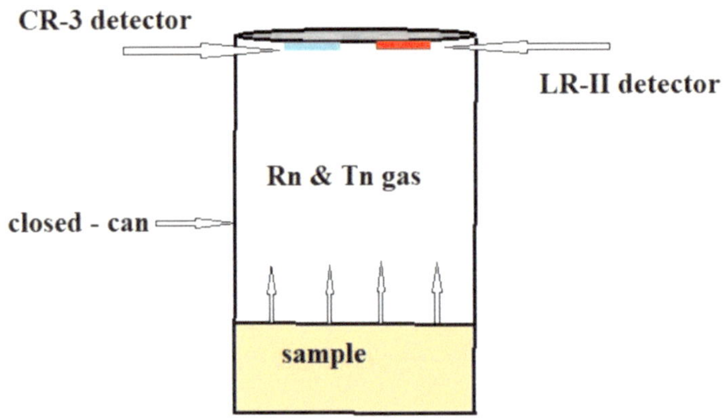

Figure 2.
Closed-can technique (two different detectors).

At radioactive secular equilibrium between radon and its progenies, i.e.:

$$\frac{C_{Po218}}{C_{Rn}} = \frac{C_{Po214}}{C_{Rn}} = 1 \tag{44}$$

Eq. (43) becomes:

$$\rho_{Rn}^{G} = K_{RnC} C_{Rn} T \tag{45}$$

where

$$K_{RnC} = K_{Rn} + K_{Po218} + K_{Po214} \tag{46}$$

By same procedure, we find ρ_{Th}^{G} for thoron on CR-39 detector. The total track density ρ_{CR} for radon and thoron become:

$$\rho_{CR} = \rho_{Rn}^{G} + \rho_{Th}^{G} = K_{RC} C_{Rn} T + K_{TC} C_{Th} T \tag{47}$$

The final result, we obtain tow equation description relationship between the radon and thoron concentration and tack density on CR-39 and LR-II detectors:

$$\left.\begin{aligned} \rho_{CR} &= K_{RC}\, T\, C_{Rn} + K_{TC}\, T C_{Th} \\ \rho_{LR} &= K_{RL}\, T\, C_{Rn} + K_{TL}\, T C_{Th} \end{aligned}\right\} \tag{48}$$

This are tow algebraic liner equations by tow variables C_{Rn} and C_{Th}, the general solution for this are:

$$\left.\begin{aligned} C_{Rn} &= \frac{K_{TL}\,\rho_{CR} - K_{TC}\rho_{LR}}{K_{RC}K_{TL} - K_{RL}K_{TC}} \\ C_{Th} &= \frac{K_{RL}\,\rho_{CR} - K_{RC}\rho_{LR}}{K_{RL}K_{TC} - K_{RC}K_{TL}} \end{aligned}\right\} \tag{49}$$

After performing a series of mathematical simplifications, we obtain mathematical relationships to calculate concentrations for radon and thoron as function to tracks density ρ_r, radon-thoron concentrations C_r and calibration factors $(K_L, K_{Rn}, K_{Th}, K_L)$ ratio.

$$\left.\begin{aligned} C_{Rn} &= \frac{\rho_{CR}}{(K_{RC} + K_{TC}\,C_r)\,T} \\ C_{Tn} &= C_r C_{Rn} \end{aligned}\right\} \tag{50}$$

$$\left.\begin{aligned} C_r &= K_L\left(\frac{\rho_r - K_{Rn}}{K_{Th} - \rho_r}\right) \\ \rho_r &= \frac{\rho_{CR}}{\rho_{LR}} \end{aligned}\right\} \tag{51}$$

$$\left.\begin{aligned} K_{Rn} &= \frac{K_{RC}}{K_{RL}} \\ K_{Th} &= \frac{K_{TC}}{K_{TL}} \\ K_L &= \frac{K_{RL}}{K_{TL}} \end{aligned}\right\} \tag{52}$$

where:
ρ_{CR}: Tracks density registration on CR-39 detector.
ρ_{LR}: Tracks density registration on LR-115(II) detector.
K_{RC}: Calibration factor of radon for CR-39 detector.
K_{RL}: Calibration factor of radon for LR-115(II) detector.
K_{TC}: Calibration factor of thoron for CR-39 detector.
K_{TL}: Calibration factor of thoron for LR-115(II) detector.
Since, C_r is positive quantity, the CR-LR track density ratio should be $K_{Rn} < \rho_r < K_{Th}$ and $K_{Th} > K_{Rn}$. Since, $\rho_{CR} > \rho_{LR}$, because CR-39 recorders all alpha particles energies, while LR-II detector recorders window energies (i.e. between E_{min} and E_{max}), this mean $\rho_r > 1$ [9].

3. Indoor radon technique measurements

Radon and thoron present in outdoor and indoor air as they exhaled from soil and building materials of the walls, floor and ceilings. It is critically important that inhalation of radon and their progeny concentrations has been shown experimentally to cause lung cancer in rats and observed to cause lung cancer in men exposed to large amounts in the air of mines. Several techniques have been developed to measure radon indoors. The use of a Solid State Nuclear Track Detector (SSNTD) closed in a cup mode (passive dosimeters) and bare mode has turned out to be the most appropriate for long term measurements. Radon measurement with, bare (open) detector are rough and with rather high uncertainties. Some device of cup mode have single chamber called invented cup. In twin cup mode, the detectors are kept inside a twin cup dosimeter, a cylindrical plastic chamber divided into two equal compartments. The two equal compartments on both sides are filter and pinhole compartments. There is one small compartment at the external middle attached to it which is used for bare mode exposure as shown in **Figure 3** [1, 10, 11].

This design of the dosimeter was well suited to discriminate ^{222}Rn and ^{220}Rn in mixed field situations, where both gases are present. SSNTDs were used as detectors and affixed at the bottom of each cup as well as on the outer surface of the dosimeter. The exposure of the detector inside the cup is termed as cup mode and the one exposed open is termed as the bare mode. One of the cups had its entry covered with a glass fiber filter paper that permits both ^{222}Rn and ^{220}Rn gases into the cup and is called the filter cup. The other cup was covered with a soft sponge and is called the sponge cup. SSNTDs film inside the sponge cup registers tracks contributed by ^{222}Rn only, while that in the filter cup records tracks due to ^{222}Rn

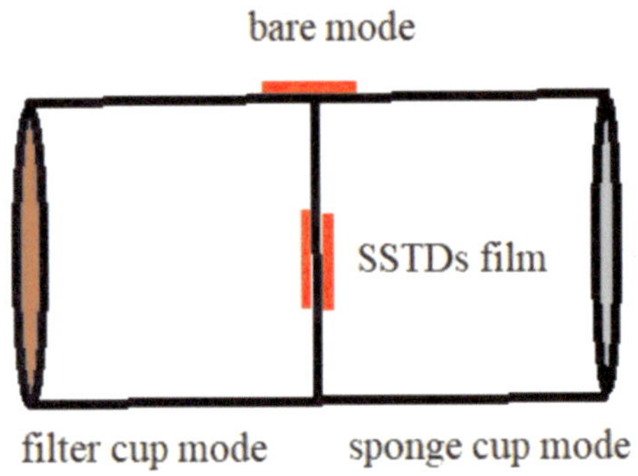

Figure 3.
Twin cups dosimeter.

and ^{220}Rn. The third SSNTDs film exposed in the bare mode registers alpha tracks contributed by the concentrations of both the gases and their alpha emitting progenies. The dosimeters were kept at a height of 1.5 m from the ceiling of the room and care was taken to keep the bare card at least 10 cm minimum away from any surface. This ensured that errors due to tracks from deposited activity from nearby surfaces were avoided, since the ranges of alpha particles from ^{222}Rn/^{220}Rn are about 10 cm. The global track density rates, due to α-particles emitted by radon, registered on the SSNTDs can write mathematical [11]:

$$\rho_S = K_{RS}\, T\, C_{Rn} \tag{53}$$

$$\rho_F = K_{RF}\, T\, C_{Rn} + K_{TF}\, T\, C_{Th} \tag{54}$$

The general solution for this are:

$$C_{Rn} = \frac{\rho_S}{K_{RS} T} \tag{55}$$

$$C_{Th} = \frac{\rho_S}{K_{TF} T}\left(\frac{\rho_F}{\rho_S} - \frac{K_{RF}}{K_{RS}}\right) \tag{56}$$

where:
ρ_S = Tracks density for sponge cup.
ρ_F = Tracks density for filter piper cup.
K_{RS} =Calibration factor of radon for sponge cup.
K_{RF} = Calibration factor of radon for filter piper cup.
K_{TF} = Calibration factor of thoron for filter piper cup.
Since, C_{Th} is positive quantity, $\frac{\rho_F}{\rho_S}$ track density ratio should be $\frac{\rho_F}{\rho_S} > \frac{K_{RF}}{K_{RS}}$.

4. Equilibrium factor

The equilibrium factor (F) is defined as the ratio of potential alpha energy concentration ($PAEC$) of actual air-radon mixture (also radon progeny) to the PAEC in secular equilibrium with radon. The equilibrium factor (F) is an important parameter in the determination of radon equivalent dose. A common practice for radon hazard assessment nowadays is to, first, determine the radon gas concentration and then to apply an assumed F with a typical value of about 0.4 [12]. However, in reality, F varies significantly with time and place, and an assumed F cannot reflect the actual conditions. Actually, the radon concentration was very dependent on the ventilation rate, therefore F was determined by SSNTD based on using can and bare method. In this method, two similar detectors were exposed to radon, one in can-mode configuration (in can detector) and the other in bare-mode configuration (bare detector). F can be found as a function of the track density ratio ρ_B/ρ_S between bare (ρ_B) and in can (ρ_S) detector, respectively. There are three steps for calculating the equilibrium factor [1, 11, 13]:

1. The potential alpha energy concentration of any mixture of (short-live) radon or thoron daughters in air.

2. The working level (WL).

3. The ventilation rate as a function of the track density ratio between bare mode and can mode.

4. The equilibrium factor as a function of the ventilation rate.

The potential alpha energy concentration ($PAEC$) of any mixture of (short-live) radon or thoron daughters in air is the sum of the potential alpha energy of all daughters atoms present per unit volume of air. The usual unit for this quantity is $\text{MeV} \cdot \text{l}^{-1}$. This unit is related to the SI units J and m^3 according to $1\ \text{J} \cdot \text{m}^{-3} = 6.24 \times 10^9\ \text{Mev} \cdot \text{l}^{-1}$ [12]. To express mathematically, we let:

$$PAEC_{Total} = PAEC_1N_1 + PAEC_2N_2 + PAEC_3N_3 + PAEC_4N_4 \tag{57}$$

where $PAEC_{Total}$, is the total potential alpha energy concentration of any mixture of (short -live) radon daughters in air. $PAEC_i$ (i = 1, 4) are the potential alpha energy for ^{218}Po, ^{214}Pb, ^{214}Bi and ^{214}Po, respectively. N_i is the number atoms of daughters of radon. The concentration potential alpha energy is defined as:

$$C_{PAET} = \frac{PAET}{V} \tag{58}$$

$$C_i = \frac{A_i}{V} = \frac{\lambda_i N_i}{V} \tag{59}$$

From Eqs. (57), (58) and (59), we get:

$$C_{PAET} = \left[\frac{PAE_1}{\lambda_1} f_1 + \frac{PAE_2}{\lambda_2} f_2 + \frac{PAE_3}{\lambda_3} f_3 + \frac{PAE_4}{\lambda_4} f_4\right] C_0 \tag{60}$$

$$\left.\begin{aligned} f_i &= \frac{C_i}{C_0} \\ C_0 &= C_{Rn} \end{aligned}\right\} \tag{61}$$

where f_i, is the activity concentration fraction. A special unit for this quantity used for radiation protection purposes is the working level (WL). A WL is defined as corresponding approximately to the potential alpha energy concentration of short-live radon daughters in air which are in radioactive equilibrium with a radon concentration of $3.7\ \text{kBq} \cdot \text{m}^{-3}$ [12].

$$C_1 = C_2 = C_3 = C_4 = C_0 = C_{Rn} = 3.7\ \text{kBq.m}^{-3} = 3.7 Bq \cdot l^{-1} \tag{62}$$

Eq. (60) become as:

$$C_{PAECT} = WL = \frac{3.7\,PE_1}{\lambda_1} + \frac{3.7\,PE_2}{\lambda_2} + \frac{3.7\,PE_3}{\lambda_3} + \frac{3.7\,PE_4}{\lambda_4} = 1.3 * 10^5\ Mev\ l^{-1} \quad (63)$$

The numerical calculations of Eq. (63) are shown in **Table 3**.

One (WL) equal to 1.3×10^5 Mev $\cdot$ l^{-1} of air. To obtain the concentration potential alpha energy by fraction energy, Eq. (63) multi and divide by 3.7 Bq $\cdot$ l^{-1} and multi and divide by 1.3×10^5 Mev $\cdot$ l^{-1}:

$$C_{PET} = \left[\frac{\frac{3.7PE_1}{\lambda_1}}{1.3 * 10^5} f_1 + \frac{\frac{3.7PE_2}{\lambda_2}}{1.3 * 10^5} f_2 + \frac{\frac{3.7PE_3}{\lambda_3}}{1.3 * 10^5} f_3 + \frac{\frac{3.7PE_4}{\lambda_4}}{1.3 * 10^5} f_4\right] \frac{1.3 * 10^5}{3.7} C_{Rn} \quad (64)$$

$$C_{PET} = \left[0.11\,f_1 + 0.51\,f_2 + 0.38\,f_3\right] \frac{1.3 * 10^5}{3.7} C_{Rn} \quad (65)$$

To obtain the concentration potential alpha energy by unit WL is:

$$C_{PET}(WL) = \frac{F_{Rn}\ C_{Rn}\left(Bq\ l^{-1}\right)}{3.7} \quad (66)$$

or

$$C_{PET}(mWL) = \frac{F_{Rn} C_{Rn}\ (Bq\ m^{-3})}{3.7} \quad (67)$$

where

$$F_{Rn} = 0.11\,f_1 + 0.51\,f_2 + 0.38\,f_3 \quad (68)$$

For Thoron daughters we find the concentration of thoron which give potential energy of alpha its daughters equal to 1.3×10^5 Mev $\cdot$ l^{-1} is 0.275 Bq $\cdot$ l^{-1}, as shown in **Table 4**.

By the same setup, we drive Eq. (60) for thoron and numerical calculations of it, as shown in **Table 5**.

The concentration potential alpha energy for thoron by unit WL is:

$$C_{PET}(mWL) = \frac{F_{Tn} C_{Tn}\left(Bq.m^{-3}\right)}{0.275} \quad (69)$$

where

$$F_{Tn} = 0.91\,f_2 + 0.09\,f_3 \quad (70)$$

Daughters of radon	α-energy (Mev)	Half-life (s)	Ultimate alpha energy (Mev)	Total energy (Mev)	Fraction energy
Po-218	6	183	6 + 7.69	1.34E+04	0.11
Pb-214	zero	1563	zero+7.68	6.41E+04	0.51
Bi-214	zero	1182	zero +7.68	4.85E+04	0.38
Po-214	7.69	1.64E-04	7.68	6.72E-03	zero
			C_{PAECT} =	1.3E+05	

Table 3.
Numerical calculations of Eq. (63).

Daughters of thoron	α-energy (Mev)	Half-life (s)	Ultimate alpha energy (Mev)	Total energy (Mev)	CTn ($Bq \cdot l^{-1}$)
Po-216	6.78	0.15	14.5952	3.16E+00	0.275
Pb-212	zero	38,304	7.8152	4.32E+05	
Bi-212	6.1	3636	7.8152	4.10E+04	
Po-212	8.78	3.04E-07	8.78	3.85E-06	
				4.73E+05	

Table 4.
Numerical calculations of thoron concentration as corresponding to the total energy 1.3×10^5 $Mev \cdot l^{-1}$.

Daughters of thoron	α-energy (Mev)	Half-life (s)	Ultimate alpha energy (Mev)	Total energy (Mev)	Fraction energy
Po-216	6.78	0.15	14.5952	8.69E-01	zero
Pb-212	zero	38,304	7.8152	1.19E+05	0.91
Bi-212	6.1	3636	7.8152	1.13E+04	0.09
Po-212	8.78	3.04E-07	8.78	1.06E-06	zero
				1.3E+05	

Table 5.
Numerical calculations of Eq. (63) for thoron.

F_{Rn} and F_{Tn} are the equilibrium factor for radon and thoron, respectively. The maximum value of equilibrium factor is $F = 1$, when the radon or thoron progeny are present in complete equilibrium with radon/thoron that is present. The minimum value of equilibrium factor is $F = 0$, that is mean no-equilibrium between the radon or thoron and its progeny. In our work, we measured the equilibrium factor depending on the ventilation rate. Ventilation rate is one of the parameters used to describe the perturbation caused in radioactive equilibrium of radon/thoron and its descendants in air. Decay of radon and production of radon can be described by the rate equations for serial radioactive decay chain (Batman equations):

$$\frac{dN_i}{dt} = \lambda_{i-1}N_{i-1} - \Lambda_i N_i \ (\mathrm{i} = 1 \dots 4) \tag{71}$$

where

$$\Lambda_i = V + A_i + W_i + \lambda_i \tag{72}$$

The first term on the right is the rate of formation of the ith-member of the progeny by radioactive decay of the (i-1)th-member, with constant λ_{i-1}; the second term describes the radioactive leakage rate, owing to ventilation V, to aerosol grains A_i and deposition on the walls W_i to which it added the rate of radioactive decay of the ith-member of the progeny, λ_i. The index i, running from 1 to 4, labels the relevant daughter in the radon family: ^{218}Po, ^{214}Pb, ^{214}Bi, ^{214}Po. ^{222}Rn itself will be label with i = 0. For thoron family: ^{216}Po, ^{212}Pb, ^{212}Bi, ^{212}Po. ^{220}Rn (Tn) itself will be labeled with i = 0. Ventilation rate affects equally all members of the family. When steady state is reached, radon daughter's activities Eq. (71) become as [13]:

$$\lambda_{i-1}N_{i-1} = \Lambda_i N_i \tag{73}$$

or

$$\frac{C_i}{C_{i-1}} = \frac{\lambda_i}{\Lambda_i} = d_i \tag{74}$$

From Eq. (74), we obtain:

$$\left.\begin{aligned} f_1 &= \tfrac{C_1}{C_0} = d_1 \\ f_2 &= \frac{C_2}{C_0} = d_1 d_2 \\ f_3 &= \frac{C_3}{C_0} = d_1 d_2 d_3 \\ f_4 &= \tfrac{C_4}{C_0} = d_1 d_2 d_3 \end{aligned}\right\} \tag{75}$$

where $\frac{C_4}{C_3} = d_4 = 1$ since $C_3 = C_4$For $\lambda_3 \ll \lambda_4$

When the ventilation rate is the only environmental affecting disequilibrium or when it is the dominant on, ventilation rates and the equilibrium factor are obtained by track density measurements, so:

$$\Lambda_i = V + \lambda_i \tag{76}$$

The track density of both bare mode and can mode (with sponge filter) detector relates the concentration of radon and its daughters as:

In can mode:

$$\rho_S = \rho_0 + \rho_1 + \rho_4 \tag{77}$$

$$\rho_S = K_0 C_0 T + K_1 C_1 T + K_4 C_4 T \tag{78}$$

$$\rho_S = (K_0 + K_1 f_1 + K_4 f_4) C_0 T \tag{79}$$

In can mode $f_1 = f_2 = f_3 = f_4 = 1$, Eq. (79) become as:

$$\rho_S = K_S C_0 T \tag{80}$$

where

$$K_s = K_0 + K_1 + K_4 \tag{81}$$

In bare mode (in absence thoron)

$$\rho_B = \rho_0 + \rho_1 + \rho_4 \tag{82}$$

$$\rho_B = (\overline{K}_0 + \overline{K}_1 f_1 + \overline{K}_4 f_4) \overline{C}_0 T \tag{83}$$

In bare mode $f_1 \neq f_2 \neq f_3 \neq f_4$ because no equilibrium between the radon and its progenies, but $\overline{K}_0 = \overline{K}_1 = \overline{K}_4 = K_B$, therefore Eq. (83) become as:

$$\rho_B = K_B (1 + f_1 + f_4) \overline{C}_0 T \tag{84}$$

By dividing Eq. (84) on Eq. (80) when $\overline{C}_0 = C_0$, we obtain:

$$1 + f_1 + f_4 = K_{SB} \rho_{BS} \tag{85}$$

$$\left.\begin{aligned} \rho_{BS} &= \frac{\rho_B}{\rho_S} \\ K_{SB} &= \frac{K_S}{K_B} \end{aligned}\right\} \tag{86}$$

From Eqs. (74), (75), (76) and (84), we obtain:

$$\frac{\lambda_1}{(\lambda_1 + V)} + \frac{\lambda_1}{(\lambda_1 + V)} \frac{\lambda_2}{(\lambda_2 + V)} \frac{\lambda_3}{(\lambda_3 + V)} = K_{SB}\rho_{BS} - 1 \tag{87}$$

After some setup, we obtain:

$$V^3 + a_2 V^2 + a_1\, V + a_0 = 0 \tag{88}$$

where

$$\left.\begin{aligned} a_2 &= (1 + \mathrm{B})\lambda_1 + \lambda_2 + \lambda_3 \\ a_1 &= (1 + \mathrm{B})(\lambda_2 + \lambda_3)\lambda_1 + \lambda_2\lambda_3 \\ a_0 &= (1 + 2\mathrm{B})\lambda_1\lambda_2\lambda_3 \\ \mathrm{B} &= \frac{1}{1 - K_{SB}\rho_{BS}} \end{aligned}\right\} \tag{89}$$

Ventilation rate is the solution of Eq. (89), obtainable by means of standard algebraic procedures. The equilibrium factor is strongly dependent on the ventilation rate. This dependence was expressed by using Eqs. (68), (70), (75) and (76).

$$F_{Rn} = \frac{\lambda_1}{(\lambda_1 + V)} \left(0.11 + 0.51 \frac{\lambda_2}{(\lambda_2 + V)} + 0.38 \frac{\lambda_2\lambda_3}{(\lambda_2 + V)(\lambda_3 + V)} \right) \tag{90}$$

$$F_{Tn} = \frac{\lambda_1\lambda_2}{(\lambda_1 + V)(\lambda_2 + V)} \left(0.91 + 0.09 \frac{\lambda_3}{(\lambda_3 + V)} \right) \tag{91}$$

Values of the equilibrium factor F, follow from solution of Eq. (88) and combined Eq. (90) and Eq. (91).

5. Determination of the radon diffusion coefficient in porous medium

The diffusion of radon is a process determined by radon gas concentration gradient across the radon gas sources (rocks, soils, building materials, and other different materials) and the surrounding air. The knowledge of radon diffusion coefficient makes it possible to determine other features of a material related to radon. The diffusion of radon through the porous medium can be described by Fick's second laws [1, 14, 15]. The basic concept in measuring radon concentration involves observations of radon diffusion through the porous medium being studied. This medium separates two chambers that have different radon concentrations. In such configurations, radon diffusion comes from one chamber to another, primarily in one direction only. The standard method for the measurement of radon diffusion coefficient beads on the evaluation of radon flux through tested material exposed to high radon concentration and placed between two containers by measuring in both sides of material. For this, one of the containers is connected to radon source (radium source) and the other where the radon concentration is periodically evaluated as shown in **Figure 4**. It is assumed that radon is well mixed in both containers.

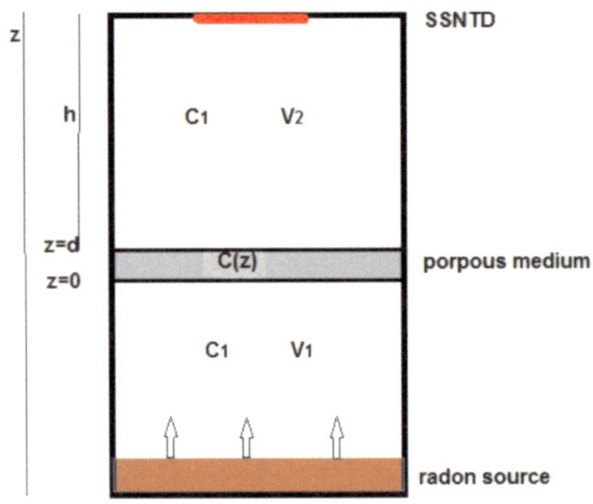

Figure 4.
Diffusion radon chambers.

The determination of the radon diffusion coefficient with this method is normally performed under steady state conditions.

The container 1 has volume V_1 and radon concentration C_1 and container 2 has volume V_2 and radon concentration C_2. It is assumed that the material represents thickness d and area A produced P is the radon production rate ($Bq \cdot m^{-3} \cdot s^{-1}$) and has diffusion coefficient D. We neglect radon production within the porous material itself and no leakage in containers. Radon transport in porous material is described by a general equation of continuity, which includes four basic processes: generation, decay, diffusion and convection. Under the supposition of the time-dependent one-dimensional differential equation (no convection) describing the radon activity concentration C_0 is given by Fick's second law:

$$\frac{\partial C_0(z,t)}{\partial t} = D_e \frac{\partial^2 C_0(z,t)}{\partial z^2} - \lambda\, C_0(z,t) + P \tag{92}$$

where $C_0(z,t)$ is the radon concentration within the pores (Bq/m^3), D is the bulk diffusion coefficient (m^2/h, gas flux expressed per unit area of material as a whole), D_e is the effective diffusion coefficient λ is the radon decay constant, ε is the prostiy and P is the production rate of radon in the pore air ($Bq/m^3 \cdot h$):

$$D_e = \frac{D}{\varepsilon} \tag{93}$$

$$P = \frac{f\, \lambda\, C_{Ra}\, \rho}{\varepsilon} \tag{94}$$

where, f is the emanation fraction, ρ is the bulk density (kg/m^3) and C_{Ra} is the radium concentration (Bq/kg).

In steady state $\frac{\partial C_0(z,t)}{\partial t} = 0$ and Eq. (92) becomes:

$$\frac{d^2 C_0(z)}{dz^2} - \frac{\lambda}{D_e} C_0(z) + \frac{P}{D_e} = 0 \tag{95}$$

The general solution is:

$$C_0(z) = A\, e^{z/l} + B\, e^{-z/l} + \frac{P}{\lambda} \tag{96}$$

$$l = \sqrt{\frac{D_e}{\lambda}} \tag{97}$$

where A, B are integral constant obtained by boundary conditions, l is the diffusion length of radon (m). The boundary conditions are formed by, it is assumed that the radon concentration at the boundary conditions is specified by the radon concentration of incoming air and the radon concentration is set to a constant value at these boundary conditions. Two boundary conditions at equilibrium rate, we needed:

$$\left.\begin{array}{l} C_0(z=0)=C_1 \\ C_0(z=d)=C_2 \end{array}\right\} \tag{98}$$

We get two equations:

$$C_1 = A + B + \frac{P}{\lambda} \tag{99}$$

$$C_2 = A\, e^{\beta} + B\, e^{-\beta} + \frac{P}{\lambda} \tag{100}$$

where

$$\beta = \frac{d}{l} \tag{101}$$

By solving Eqs. (99) and (100) and some mathematical step, we get:

$$A = \frac{-e^{-\beta}}{e^{\beta} - e^{-\beta}} C_1 + \frac{1}{e^{\beta} - e^{-\beta}} C_2 - \frac{1 - e^{-\beta}}{e^{\beta} - e^{-\beta}} \frac{P}{\lambda} \tag{102}$$

$$B = \frac{e^{\beta}}{e^{\beta} - e^{-\beta}} C_1 - \frac{1}{e^{\beta} - e^{-\beta}} C_2 + \frac{1 - e^{\beta}}{e^{\beta} - e^{-\beta}} \frac{P}{\lambda} \tag{103}$$

By using the relations:

$$\left.\begin{array}{l} e^{\beta} + e^{-\beta} = 2\cosh\beta \\ e^{\beta} - e^{-\beta} = 2\sinh\beta \end{array}\right\} \tag{104}$$

And mathematical steps, we get the general solution:

$$C_0(z) = \frac{\sinh\left(\frac{d-z}{l}\right)}{\sinh\left(\frac{d}{l}\right)} C_1 + \frac{\sinh\left(\frac{z}{l}\right)}{\sinh\left(\frac{d}{l}\right)} C_2 + \left(\frac{\sinh\left(\frac{d-z}{l}\right) + \sinh\left(\frac{z}{l}\right)}{\sinh\left(\frac{d}{l}\right)} - 1\right) \frac{P}{\lambda} \tag{105}$$

From steady state of exhalation radon Eq. (29), we get:

$$E = \frac{\lambda\, V\, C_2}{A} \tag{106}$$

From Fick's first law:

$$E = -D_e \left.\frac{dC_0(z)}{dz}\right|_{z=d} \tag{107}$$

We get:

$$\frac{dC_0(z)}{dz} = \frac{-\cosh\left(\frac{d-z}{l}\right)}{l \sinh\left(\frac{d}{l}\right)} C_1 + \frac{\cosh\left(\frac{z}{l}\right)}{l \sinh\left(\frac{d}{l}\right)} C_2 + \left(\frac{\cosh\left(\frac{z}{\alpha}\right) - \cosh\left(\frac{d-z}{\alpha}\right)}{l \sinh\left(\frac{d}{l}\right)}\right) \frac{P}{\lambda} \tag{108}$$

$$\left.\frac{dC_0(z)}{dz}\right|_{z=d} = \frac{-1}{l\sinh\left(\frac{d}{l}\right)}C_1 + \frac{\cosh\left(\frac{d}{l}\right)}{l\sinh\left(\frac{d}{l}\right)}C_2 + \left(\frac{\cosh\left(\frac{d}{\alpha}\right)-1}{l\sinh\left(\frac{d}{l}\right)}\right)\frac{P}{\lambda} \tag{109}$$

Substitute Eq. (109) in Eq. (107), we obtain:

$$E = \frac{-D_e}{l\ \sinh\left(\frac{d}{l}\right)}\left\{-C_1 + \cosh\left(\frac{d}{l}\right)C_2 - \left(\cosh\left(\frac{d}{l}\right)-1\right)\frac{P}{\lambda}\right\} \tag{110}$$

From Eq. (106) and Eq. (110), we get:

$$C_2 = \frac{C_1}{\cosh\left(\frac{d}{l}\right) + \frac{\lambda l V_2}{D_e A}\ \sinh\left(\frac{d}{l}\right)} + \frac{1-\ \cosh\left(\frac{d}{l}\right)}{\cosh\left(\frac{d}{l}\right) + \frac{\lambda l V_2}{D_e A}\ \sinh\left(\frac{d}{l}\right)}\frac{P}{\lambda} \tag{111}$$

Since,

$$\left.\begin{aligned} V_2 &= A\,h \\ l^2 &= \frac{D_e}{\lambda} \\ \frac{\lambda l V_2}{D_e A} &= \frac{h}{l} \end{aligned}\right\} \tag{112}$$

Substitute set Eq. (112) in Eq. (111), we get:

$$C_2 = \frac{C_1}{\cosh\left(\frac{d}{l}\right) + \frac{h}{l}\ \sinh\left(\frac{d}{l}\right)} + \frac{1-\ \cosh\left(\frac{d}{l}\right)}{\cosh\left(\frac{d}{l}\right) + \frac{h}{l}\ \sinh\left(\frac{d}{l}\right)}\frac{P}{\lambda} \tag{113}$$

Rewrite Eq. (113):

$$\left(1-\frac{P}{\lambda\,C_2}\right)\cosh\left(\frac{d}{l}\right) + \frac{h}{l}\ \sinh\left(\frac{d}{l}\right) - \left(\frac{C_1}{C_2} - \frac{P}{\lambda\,C_2}\right) = 0 \tag{114}$$

From Eqs. (101) and (114), we get final relation:

$$\left(1-\frac{P}{\lambda\,C_2}\right)\cosh\beta + \frac{h}{d}\beta\sinh\beta - \left(\frac{C_1}{C_2} - \frac{P}{\lambda\,C_2}\right) = 0 \tag{115}$$

Eq. (115) is nonlinear equation; therefore, the radon diffusion coefficient D can be numerically calculated by using the Newton-Raphson method.

IF β is small, and by using set relation in Eq. (116), we can make approximation to evolution radon diffusion coefficient by simple relation as:

$$\left.\begin{aligned} \sinh\beta &= \frac{e^{\beta}-e^{-\beta}}{2} = \frac{1+\beta-1+\beta}{2} = \beta \\ \cosh\beta &= \frac{e^{\beta}+e^{-\beta}}{2} = \frac{1+\beta+1-\beta}{2} = 1 \end{aligned}\right\} \tag{116}$$

We find simple relation to calculate the effective diffusion coefficient Eq. (117).

$$D_e = \frac{\lambda\,d\,h}{\frac{C}{C_2} - 1} \tag{117}$$

6. Transmission factor

The purpose of using covers of cup (sponge or paper filter) to allow entry radon and thoron gases only without its daughters for filter paper and radon only for

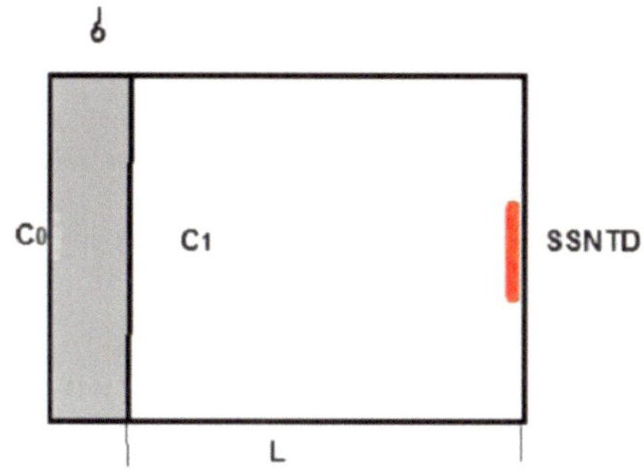

Figure 5.
Diffusion radon cup.

sponge. In actual terms it means that the covers are reduced from concentration inter cup to detection [16]. Therefore, we define transmission factor as the ratio between the concentrations inter cup and outside cup. The purpose from calculate diffusion coefficient and transmission factor to get actual radon concentration in air atmosphere. It is assumed that the gas enters the chamber through porous filter by the process of diffusion with diffusion coefficient D_1 and D_2 in air. If C_1 is the average radon/thoron gas concentration in the cup and C_o is out the cup. Let the radon diffusion in x-direction **Figure 5** [1, 14, 16].

Then, the steady state equations for described the radon gas diffusion internal porous filter and in air space in cup may be written as Eq. (95), ($P = 0$, because of the lack of a source of radium) respectively:

$$\frac{d^2C_1(x)}{dx^2} - \frac{\lambda}{D_1}C_1(x) = 0 \tag{118}$$

$$\frac{d^2C_2(x)}{dx^2} - \frac{\lambda}{D_2}C_2(x) = 0 \tag{119}$$

The general solutions are

$$C_1(x) = A_1 e^{x/l_1} + B_1 e^{-x/l_1} \tag{120}$$

$$C_2(x) = A_2 e^{x/l_2} + B_2 e^{-x/l_2} \tag{121}$$

$$\left. \begin{aligned} l_1 &= \sqrt{\frac{D_1}{\lambda}} \\ l_2 &= \sqrt{\frac{D_2}{\lambda}} \end{aligned} \right\} \tag{122}$$

where A_1, B_1, A_2, B_2, are integral constants obtained by boundary conditions, l_1, l_2 are the diffusion length of radon (m), D_1, D_2 are diffusion coefficients for porous filter and air respectively.

1. The boundary conditions at equilibrium at sides of porous filter are:

$$\left. \begin{aligned} &A_1 e^{\delta/l_1} + B_2 e^{-\delta/l_1} = C_o \qquad at \quad x = -\delta \\ &A_1 + B_1 = C_i \qquad at \quad x = 0 \end{aligned} \right\} \tag{123}$$

The solutions of these equations are:

$$\left.\begin{aligned} A_1 &= \frac{e^{\delta/l_1}}{e^{\delta/l_1}-e^{-\delta/l_1}} C_i - \frac{1}{e^{\delta/l_1}-e^{-\delta/l_1}} C_o \\ B_1 &= \frac{-e^{-\delta/l_1}}{e^{\delta/l_1}-e^{-\delta/l_1}} C_i + \frac{1}{e^{\delta/l_1}-e^{-\delta/l_1}} C_o \end{aligned}\right\} \quad (124)$$

From Eqs. (120) and (124), we get the general solution is:

$$C_1(x) = \frac{\sinh\left(\frac{x-\delta}{l_1}\right)}{\sinh\left(\frac{\delta}{l_1}\right)} C_i + \frac{\sinh\left(\frac{x}{l_1}\right)}{\sinh\left(\frac{\delta}{l_1}\right)} C_o \quad (125)$$

2. The boundary conditions at equilibrium in air space enter cup are:

$$\left.\begin{aligned} &C_2(x=0) = C_i \\ &\left.\frac{dC_2(x)}{dx}\right|_{x=L} = 0 \end{aligned}\right\} \quad (126)$$

We get two equations:

$$\left.\begin{aligned} &A_2 + B_2 = C_i && at && x = 0 \\ &\frac{A_2}{l_2} e^{L/l_2} - \frac{B_2}{l_2} e^{-L/l_2} = 0 && at && \left.\frac{dC_2(x)}{dx}\right|_{x=L} = 0 \end{aligned}\right\} \quad (127)$$

The solutions of these equations are:

$$\left.\begin{aligned} A_2 &= \frac{e^{-L/l_2}}{e^{L/l_2} + e^{-L/l_2}} C_i \\ B_2 &= \frac{e^{L/l_2}}{e^{L/l_2} + e^{-L/l_2}} C_i \end{aligned}\right\} \quad (128)$$

From Eqs. (121) and (128), we get the general solution as:

$$C_2(x) = \frac{\sinh\left(\frac{x-L}{l_2}\right)}{\sinh\left(\frac{L}{l_2}\right)} C_i \quad (129)$$

The average concentration radon inter cup, we get from relation:

$$\overline{C} = \frac{1}{L} \int_0^L C_2(x) dx \quad (130)$$

The average concentration radon inter cup:

$$\overline{C} = \frac{l_2}{L} \tanh\left(\frac{L}{l_2}\right) C_i \quad (131)$$

3. The boundary condition number 3 is:

$$E_1|_{x=0} = E_2|_{x=0} \quad (132)$$

or

$$\left.\begin{array}{l} -D_1 \left.\dfrac{dC_1(x)}{dx}\right|_{x=0} = -D_2 \left.\dfrac{dC_2(x)}{dx}\right|_{x=0} \\ \left.\dfrac{dC_1(x)}{dx}\right|_{x=0} = \dfrac{1}{l_1}\left\{\dfrac{\cosh\left(\frac{\delta}{l_1}\right)}{\sinh\left(\frac{\delta}{l_1}\right)} C_i - \dfrac{1}{\sinh\left(\frac{\delta}{l_1}\right)} C_o\right\} \end{array}\right\} \tag{133}$$

and

$$\left.\frac{dC_2(x)}{dx}\right|_{x=0} = -\frac{1}{l_2}\tanh\left(\frac{L}{l_2}\right)C_i \tag{134}$$

Substitute Eq. (134) in Eq. (133), we get:

$$C_i = \frac{C_o}{\cosh\left(\frac{\delta}{l_1}\right) + \frac{D_2 l_1}{D_1 l_2}\sinh\left(\frac{\delta}{l_1}\right)\tanh\left(\frac{L}{l_2}\right)} \tag{135}$$

IF $\frac{\delta}{l_1}$ is small $\sinh\left(\frac{\delta}{l_1}\right) = \frac{\delta}{l_1}$ and $\cosh\left(\frac{\delta}{l_1}\right) = 1$, we get:

$$\overline{C} = \frac{\frac{l_2}{L}}{\frac{\lambda\, l_2 \delta}{D_1} + \coth\left(\frac{L}{l_2}\right)} C_o \tag{136}$$

From define transmission factor we get as:

$$Transmission\ (\%) = \frac{\overline{C}}{C_o} = \frac{\frac{l_2}{L}}{\frac{\lambda\, l_2 \delta}{D_1} + \coth\left(\frac{L}{l_2}\right)} \tag{137}$$

7. The calibration factor

Quantitative measurements with both single and multi-SSNT detectors device can be performed only if the calibration factor is known. It can be measured experimentally or estimated theoretically. Moreover, the theoretically derived calibration factor relations may supply reasonable basic design criteria. There are many method to deriving formulas of calibration factor such as Mont- Carlo, mean critical angle by unit cm. At the same time, the radon chamber is used to determine the calibration factors for different dosimeter geometry configuration [1, 17–19]. The theoretical calibration factor is calculated by unit (cm), while the experiment calibration factor is measured by $\left(\frac{Track\ cm^{-2}}{Bq \cdot m^{-3} \cdot day}\right)$. Since 1 Bq = Disintegration/second = track/second, 1 day = 86,400 sec and 1 m^3 = 10^6 cm^3, so, 1 cm = 0.0864 $\left(\frac{Track\ cm^{-2}}{Bq \cdot m^{-3} \cdot day}\right)$ [20]. The relation between $K_{theoretical}$ and $K_{experement}$ in the can technique is:

$$K_{experement}\left(\frac{Track\ cm^{-2}}{Bq \cdot m^{-3} \cdot day}\right) = 0.0864 \sum_i K_i^{theoretical}(cm) \tag{138}$$

8. Conclusion remarks

1. The concentration of radon emanated from each sample inside the closed container reach to the steady state after a fixed period of time without leakage out of the container Eq. (10).

2. The detector should be exposed for at least 27 day to record tracks for radon concentration.

3. The radon concentration reaches the equilibrium state at the same time ($t \cong 7T_{(1/2)}$) regardless of the volume of the container.

4. There are three different quantities measured by track density, integrated radon concentration, average radon concentration and saturate radon concentration as shown in **Table 1**.

5. From the relations in **Table 2**, we get some relationship between radon concentration, area and mass radon exhalation and effective radium content are liner equations **Table 2**.

6. The radon/thoron concentrations equations which used in technique two different SSNTDs are very critical because it's a mathematical relationship as a fraction.

7. For more accurate estimation of the effective dose from radon/thoron exposure, one should measure the equilibrium factor at each site.

8. High dose does not necessarily mean there is a high concentration of radon, there may be high equilibrium factor (bad ventilation), and so we recommend interest in ventilation factor when houses and buildings design.

9. It is necessary to correct the radon concentration when using covers (sponge or filter paper) in measurements of radon concentration by transmission factor.

Acknowledgements

I would like to express my deep appreciation to:

- My supervisor Prof. Dr. Abdul Ridha Hussein Subber
- Department of physics
- Education College for Pure Science, Basra University
- Oil Ministry, Gas Filling Company and My Company (Basra Oil)
- Dr. Basim Almayahi and IntechOpen publishing

My great thanks to all which they could assistance me in any way.

DOI: http://dx.doi.org/10.5772/intechopen.92647

Author details

Ali Farhan Nader Alrekabi
Iraq Oil Ministry/Basra Oil Company, Basra University, Basra, Iraq

*Address all correspondence to: alialrekabi251@gmail.com

References

[1] Nader AF. Theoretical and Experimental Study to Evaluate Radioactivity Applied on a Selected Area in Basra Governorate [thesis]. Basra, Iraq: University of Basra, Iraq; 2015

[2] Abd-Elzaher M. An overview on studying 222Rn exhalation rates using passive technique solid-state nuclear track detectors. American Journal of Applied Sciences. 2012;**9**(10):1653-1659

[3] Tufail M, Mirza SM, Mahmood A, Qureshi AA. Application of a closed-can technique for measuring radon exhalation from mine samples of Punjab, Pakistan. Journal of Environmental Radioactivity. 2000;**50**: 267-275

[4] Jiang H, Liangquan GE, Lin Y, Yi GU. Preliminary study on a regional radon concentration in surface soil prediction method. Progress in Nuclear Science and Technology. 2011;**1**:364-367

[5] Byun SH. Radioactivity: Radioisotopes and Radiation Methodology Lecture Notes. Canada: University of McMaster; 2014

[6] Saad AF, Abdalla YK, Hussein NA, Elyaseery IS. Radon exhalation rate from building material used on the Garyounis. University campus, Benghazi Libya,Turkish. Journal of Environmental Science and Engineering. 2010;**34**:67-74

[7] Eman SA, Nageeb SH, El-Sersy AR. U and Th determination in natural samples using CR-39 and LR-115 track detectors. World Journal of Nuclear Science and Technology. 2012;**2**:36-40

[8] Koo VSY, Yip CWY, Ho JPY, Nikezik D, Yu KN. Sensitivity of LR- 115 detector in diffusion chamber to Rn^{222} in the presence of Rn^{220}. Applied Radiation and Isotopes. 2002;**56**:953-956

[9] Yu KN, Nikezic D. Long-Term measurements of radon progeny concentration with solid-state nuclear track detectors. Radiation Measurements. 2005;**40**:560-568

[10] Somogyi G, Paripas B, Varga Z. Measurement of radon, radon daughters and thoron concentration by multi-detector devices. Nuclear Tracks and Radiation Measurements. 1984;**8**(1–4): 423-427

[11] Nader AF. The determination of equilibrium factor of radon and thoron using LR-115 type II detector in a selected area from Basra governorate, Iraq. Journal of Physics Conference Series. 2019;**1258**:012032

[12] UNSCEAR. United Nations scientific committee on the effects of atomic radiation, annex D, distribution. Applied Radiation and Isotopes. 1982; **54**:467-473

[13] Abo-Elmaged M, Mansy M, Eissa HM, El-Fiki MA. Major parameters affecting the calculation of equilibrium factor using SSNTD-measured track densities. Radiation Measurements. 2006;**41**:235-240

[14] Hosoda M, Tokonami S, Sorimachi A, Janik M, Ishikaw T, Yatabe Y, et al. Experimental system to evaluate the effective diffusion coefficient of radon. Review of Scientific Instruments. 2009;**80**:013501

[15] Nader AF, Juber JH, Subber ARH, Al-Hashimi NHN. The diffusion coefficient and calibration factors for indoor radon measurement in bare and twin cup modes. Journal of Zankoy Sulaimani-A. 2016;**18**(1):49-56

[16] Eappen KP, Sahoo BK, Ramachandran TV, Mayya YS. Calibration factor for thoron estimation

in cup dosimeter. Radiation Measurements. 2008;**43**:418-421

[17] Nader AF, Subber ARH, Al-Hashimi NHN. Numerical and analytical calculations of efficiency and calibration factor for CR-39 detectors in the chamber diffusion by using Monte-Carlo method and the mean critical angle. Scholars Research Library. 2014; **5**:23-30

[18] Nader AF, Subber ARH, Al-Hashimi NHN. Mathematical calculation to find the best chamber and detector radii used for measurements the range of α-particle. The International Journal of Engineering & Science (THE IJES). 2014;**3**:48-52

[19] Subber ARH, Al-Hashmi NHN, Nader AF, Jebur JH, Khodier MK. Construct as a simple radon chamber for measurement of radon detectors calibration factors. Pelagia Research Library. 2015;**6**:128-131

[20] Ismail AH, Jaafar MS. Experimental measurement on CR-39 response for radon gas and estimation the optimum dimension of dosimeters for detection of radon. In: Proceeding of the 3ed Asian Physics Symposinum. Bandung, Indonesia; 2009

Section 3

Applications in Ionizing Radiation

Chapter 7

Production of the ^{103}Pd via Cyclotron and Preparation of the Brachytherapy Seed

Pooneh Saidi and Mahdi Sadeghi

Abstract

This study will briefly explain the production of ^{103}Pd via cyclotron for brachytherapy use. The excitation functions of ^{103}Rh(p,n)^{103}Pd and ^{103}Rh(d,2n)^{103}Pd reactions were calculated using ALICE/91, ALICE/ASH, and TALYS-1.2 codes and compared with published data. Production of ^{103}Pd was done via ^{103}Rh(p,n)^{103}Pd nuclear reaction. The target was bombarded with 18 MeV protons at 200 μA beam current for 15 h. After irradiation and radiochemical separation of the electroplated rhodium target, at the optimum condition, ^{103}Pd was absorbed into Amberlite®IR-93 resin. The preparation of the brachytherapy seed, which is loaded by the resin beads, has also been presented. At least, the method to determine the dosimetric parameters for the seed by experimental measurement has been presented.

Keywords: brachytherapy, cyclotron, production, cross-section, excitation function, palladium, rhodium

1. Introduction

Cyclotrons are charged particle circular accelerators. They are a type of particle accelerator that has many applications in nuclear physics, industry, technology, and medicine. They play an important role in medicine; for example, they are used for radiation therapy, production of medical radioisotopes, and biomedical research [1]. As a particle accelerator, one of the important uses of the cyclotron in medicine is radioisotope production [2].

For a long period, radioisotope production is basically done in nuclear reactors, but their availability is slowly decreasing, and due to some advantages of radioisotope production with the cyclotron, the development of particle accelerators started in the past century, so accelerator-based production facilities are growing, and various radioisotopes suitable for medical applications are produced.

In this chapter, the production method for the radioisotope, Palladium-103 (^{103}Pd), via cyclotron is discussed. Palladium-103 with energy emission about 20 keV results in the rapid dose falloff with the distance which is suitable for low-dose-rate (LDR) brachytherapy [3]. For nearly 25 years, brachytherapy sources containing ^{103}Pd have been clinically introduced and are in use [4, 5]. Sources containing ^{103}Pd are most commonly used in the treatment of prostate and eye cancer [6–8].

Radionuclide	$T_{1/2}$	Mode of decay	The energy of the emitted particle (keV)
^{32}P	14.26 d	$\beta^-(100)$	690
^{137}Cs	30.04 y	$\beta^-(100)$	662
^{198}Ir	73.8	$\beta^-(95.34)$	380
^{198}Au	2.69	$\beta^-(100)$	412
^{125}I	59.4 d	EC (100)	28
^{169}Yb	32.02 d	EC (100)	93
^{103}Pd	16.99 d	EC (100)	21
^{153}Sm	46.28 h	$\beta^-(100)$	223
^{142}Pr	19.12	$\beta^-(96.3)$	809
^{170}Tm	128.6 d	$\beta^-(99.87)$	66.39

Table 1.
Examples of radioisotopes commonly used in brachytherapy.

Brachytherapy is a form of treatment where a sealed radioactive source placed on or in the tissue/tumor is to be irradiated. With this method, a high dose can be delivered to the tumor with a rapid decrease in dose in the surrounding normal tissue. Brachytherapy sources are usually encapsulated; the encapsulated sources are placed within the body cavities close to the tumor, in a lumen of organs, or implanted into the tumor or placed over the tissue to be radiated.

Depending on the source dose rate, brachytherapy sources are classified into three categories [9]:

- High-dose-rate sources (HDR): >12 Gy/h radioisotopes with high-energy photon emitters like ^{137}Cs, ^{60}Co, ^{192}Ir, and ^{198}Au are used.

- Medium-dose-rate sources (MDR): 2–12 Gy/h radioisotopes are used; this category is not commonly used in brachytherapy.

- Low-dose-rate sources (LDR): Less than 2 Gy/h radioisotopes with low-energy photon emitters like ^{103}Pd and ^{125}I are used.

Table 1 shows examples of radioisotopes commonly used in brachytherapy for the three mentioned categories [10].

This study will briefly explain the production of the ^{103}Pd via cyclotron and will shortly describe some detail of ^{103}Pd production such as production process, targetry, radiochemical separation, seed fabrication, and seed dosimetry method.

2. Materials and methods

2.1 Cyclotron production of ^{103}Pd

There are different methods for the production of ^{103}Pd via cyclotron. Two of them are $^{103}Rh(p,n)^{103}Pd$ and $^{103}Rh(d,2n)^{103}Pd$ [3]. In this study, the excitation functions for both reactions have been calculated, and the optimum condition for each reaction has been obtained. Experimental data for the proton bombardment on rhodium metal as the target, via $^{103}Rh(p,n)^{103}Pd$ reaction, has also been measured. Rhodium target has been bombarded by proton in a cyclotron (Cyclone-30, IBA) with 18 MeV energy and a beam current intensity of 200 μA at the Agricultural, Medical and Industrial Research School (AMIRS) [11].

To achieve the ^{103}Pd from the irradiated target, the radiochemical separation stage is started. The problem in this stage is the dissolution of target material due to the extremely low chemical reactivity of rhodium metal. The other problem is the high quantity of rhodium in the solution. A well-known palladium extractor is dimethylglyoxime, but to prevent the decrease of extraction yield, α-furyldioxime has been used [12]. Purely obtained ^{103}Pd is then absorbed into resin; the active resins are encapsulated inside the titanium casing.

2.2 Calculation of excitation function

Excitation functions of the ^{103}Rh(p,n)^{103}Pd and ^{103}Rh(d,2n)^{103}Pd reactions were calculated using ALICE/ASH, EMPIRE (version 3.1 Rivoli) and TALYS-1.2 nuclear codes, and the TENDL-2010. Using the codes simultaneously increases the accuracy of calculations. The calculated results were compared to the existing data of references [13–18].

2.3 Nuclear models applied for cross-section calculations

The ALICE/ASH code: This code is a modified version of the ALICE code, and to describe the pre-equilibrium particle emission from nuclei, the geometry-dependent hybrid model (GDH) is used. Calculations were carried out based on the Fermi gas model with the nuclear level density parameter $a = A/y$ and the generalized superfluid nuclear model. The default value of y is equal to [17, 19].

The TALYS code: TALYS is a computer code developed at NRG Petten and CEA to predict and analyze the nuclear reactions. TALYS models the nuclear reactions that involve protons, deuterons, neutrons, alpha particles, gamma rays, tritons, and hellions. The code simulated the reactions in the energy range from 1 keV to 200 MeV, for target nuclides of mass 12 and heavier [14, 17].

EMPIRE: EMPIRE (version 3.1 Rivoli) simulates various nuclear reactions, over a broad range of energies and incident particles. This system can be used for nuclear data evaluation and also for the theoretical calculation of nuclear reactions. A projectile can be a photon, a nucleon, and light or heavy ion. There is a broad range of energy in the system; the energy range starts just above the resonance region in the case of a neutron projectile and extends up to a few hundred MeV for heavy ion-induced reactions [20].

2.4 The thickness of the target

The required thickness of the target has been calculated via the stopping and range of ions in matter (SRIM) code [21]. Based on the code results, to take full advantage of the excitation function and also to avoid the production of the radioisotope impurity, the entrance energy of the proton should be 18 MeV. The physical thickness of the rhodium layer is chosen in such a way that for a given beam/target angle geometry, the particle exit energy should be 6 MeV. The thickness of the rhodium target has to be 475 μm for 90° geometry. To minimize the thickness of the rhodium layer (and hence lower the cost price per target), a 6° geometry is preferred; in this case, the thickness of the target decreases, and a 45–50 μm layer is sufficient.

Identification of the gamma ray emitting from the radionuclides is performed by using gamma-ray spectroscopy with a high-purity germanium HP(Ge) detector (Canberra™ model GC1020-7500SL).

2.5 Preparation of ^{103}Pd brachytherapy seeds

After irradiation, the radiochemical separation phase has been started. In this phase, the $PdCl_2$ solution has been separated from rhodium, zinc, and copper.

According to the brachytherapy seed model (in case of using a sphere made of resin), resin beads and marker are encapsulated inside the titanium capsule. The end caps of the capsule are welded precisely to prevent source leakage [22]. Regarding the physical design and configuration of the source internal component, two types of designs have been used: (a) rod/wire/cylinder made of ceramic, glass, or high-Z materials and (b) sphere made of resin. In this study sphere design of the source has been discussed.

Figure 1 shows a schematic diagram of eight different brachytherapy seeds which are designed at the Agricultural, Medical, and Industrial Research School [7, 23–25].

2.6 Dosimetry of the seed

According to the American Association of Physicists in Medicine (AAPM) Radiation Therapy Committee recommendation, the dosimetry characteristics

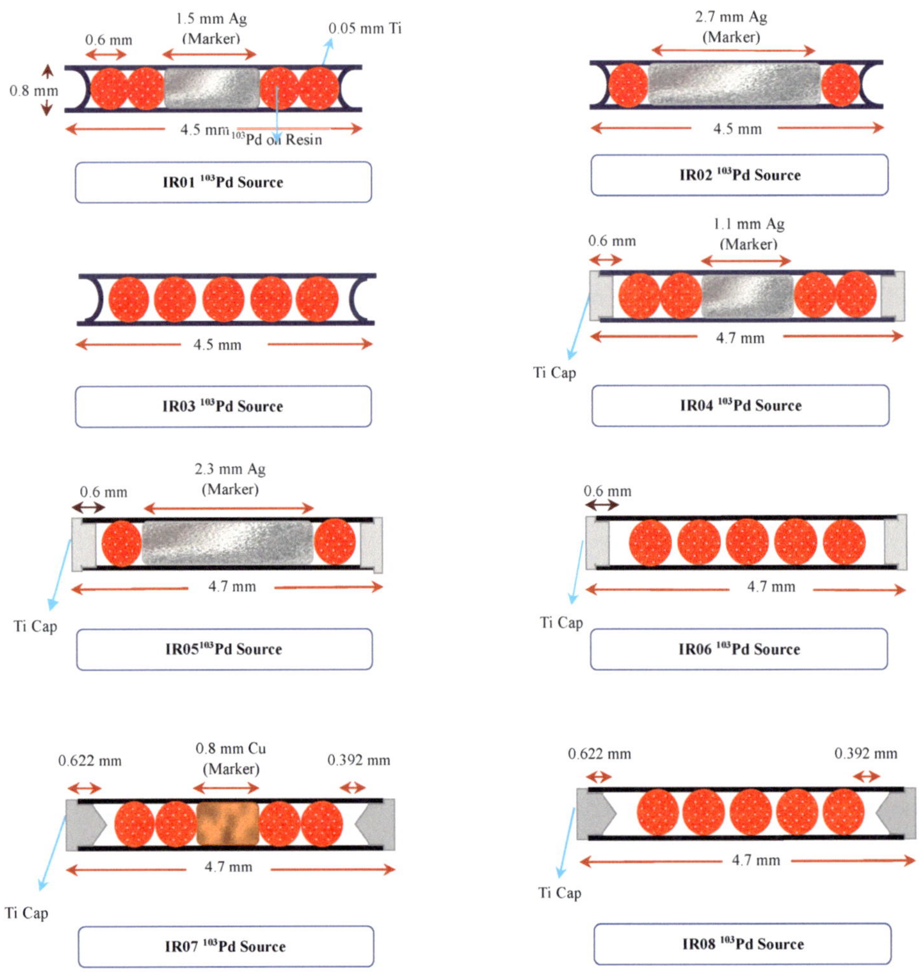

Figure 1.
Schematic drawing of the designed ^{103}Pd sources.

for all new interstitial brachytherapy seeds with energies less than 50 keV should be investigated by two independent investigators, theoretical calculations and experimental measurements. This work presents the method for thermoluminescent dosimeter (TLD) measurements to determine the dosimetric characteristics of the brachytherapy seed containing resin beads. The TLD-GR200A thermoluminescent dosimeters and two Perspex phantoms have been used, one Perspex phantom for the anisotropy function, $F(r, \theta)$, and the other for the radial dose function, $g_L(r)$.

3. Results and discussion

3.1 Excitation function

3.1.1 Excitation function study of ^{103}Rh(p,n)^{103}Pd reaction

The evaluation of the acquired data from the codes showed that the best range of the energy for proton in the ^{103}Rh(p,n)^{103}Pd reaction is 18–8 MeV. The maximum cross-section by EMPIRE (version 3.1 Rivoli) code is at E_p = 10 MeV, and the value is 574.44 mb. To evaluate the obtained results, **Figure 2** shows the comparison between the calculated results in this study and measured data by others. For the ^{103}Rh(p,n)^{103}Pd reaction, there are five cross-section measurements that exist in the literature by authors of references [26–29].

The calculated results from the TENDL-201, TALYS-1.2, and ALICE/ASH codes are in acceptable agreement with the measured data from [29], and calculated results from EMPIRE (version 3.1 Rivoli) code are in good agreement with the Hermanne et al. measured data (**Figure 2**).

3.1.2 Excitation function study of ^{103}Rh(d,2n)^{103}Pd reaction

According to the results from the codes, the optimum range of energy of the deuteron particle to produce ^{103}Pd from ^{103}Rh target for the ^{103}Rh(d,2n)^{103}Pd reaction is 22 to 8 MeV. The obtained results from ALICE/ASH hybrid model (a = A/9)

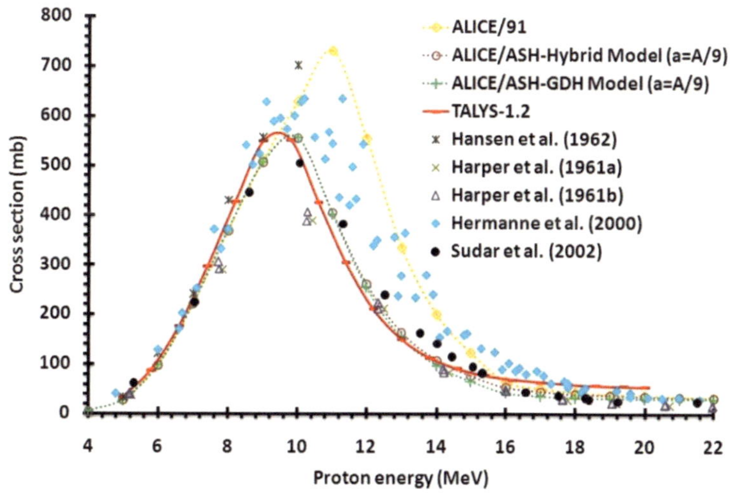

Figure 2.
Excitation function of ^{103}Rh(p,n)^{103}Pd reaction by ALICE/91, ALICE/ASH and TALYS-1.2 codes, and experimental data [10].

show that the maximum cross-section is 1158.795 mb (Ed = 13 MeV). Comparison between the calculated results of this study and the measured data obtained by [30, 31] is presented in **Figure 3**.

The obtained results from ALICE/ASH code are in good agreement with the measured data by Hermanne et al. up to 20 MeV, whereas TALYS-1.2 and EMPIRE (version 3.1 Rivoli) calculated results have lower values than ALICE/ASH results and also experimental data.

3.2 Production of ^{103}Pd

To prepare the rhodium target for irradiation, via the electrodeposition process, a thick layer of rhodium has been placed of the copper backing. According to Sadeghi et al. study [11, 12], the following conditions are the optimum conditions for the electrodeposition:

- 4.8 g rhodium (as $Rh_2(SO_4)_3$)
- pH = 2
- DC current density of 8.5 mA cm^{-2}
- 1% sulfamic acid (w/v)
- Temperature 40–60°C

After the electrodeposition process, the adhesion quality of the rhodium layer on the target backing has been tested by the thermal shock. The thermal shock has been carried out by heating the target up to 500°C for 1 h (The temperature that the Rh layer experience during high current irradiation). Thereafter, the hot target is submerged in cold water in a temperature of about 15°C. Observation of neither crack formation nor peeling off of the rhodium layers indicated a good adhesion for the purpose.

Afterward, the rhodium target was bombarded with 18 MeV protons at 200 μA beam current for 15 h (3000 μAh) [12]. At the end of the bombardment (EOB),

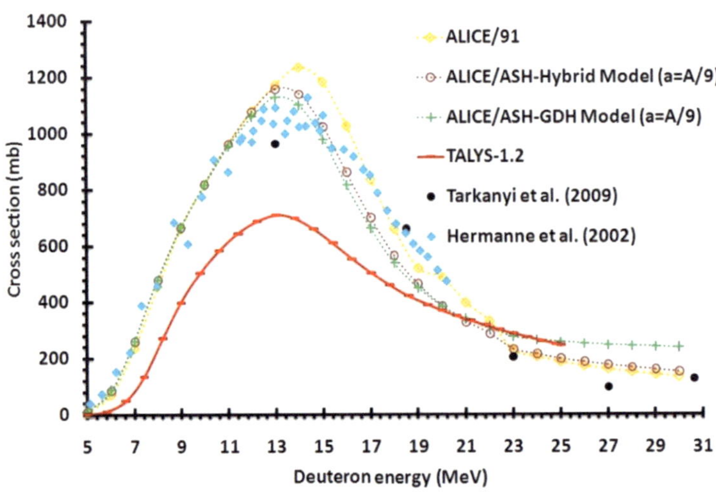

Figure 3.
Excitation function of ^{103}Rh(d,2n)^{103}Pd reaction by ALICE/91, ALICE/ASH and TALYS-1.2 codes, and experimental data [10].

the activity of ^{103}Pd and the yield of production are 685 mCi and 8.44 MBq/lAh, respectively. After the proton bombardment, the dissolution process has been started.

The optimum conditions of the electro-dissolution are as follows:

- 12 N HCl solution
- AC current density > 1.8 A cm^2
- Temperature: 75°C

After the dissolution of the target from copper backing, the residual contains $PdCl_2$, rhodium, zinc, and copper, so during the radiochemical separation phase, the $PdCl_2$ solution should be separated from rhodium, zinc, and copper. According to the data in **Figure 4**, the purity of the obtained radio-palladium is about 99%. Thereafter, the obtained filtrate solution was loaded onto the column of size Ø 0.5 cm × 2 cm packed with Amberlite®IR-93 resin with 0.6 mm diameter. The summarized results in **Figure 5** show that 0.05 M HCl is the most suitable concentration for adsorption of ^{103}Pd on the Amberlite®IR-93 resin.

The dosimetric parameters of the seeds have been determined by theoretical calculation and experimental measurement, according to TG-43 U1 report. The theoretical method to obtain the dosimetric parameters of the brachytherapy seeds has been discussed and explained in Refs. [6, 9].

The following is the method to determine the dosimetric parameters by experimental measurements.

3.3 Dosimetry method

3.3.1 Thermoluminescent dosimeters

The TLD-GR200A (PTW, Freiburg, Germany) circular chips [26] of the following specifications have been used in this study:

- 0.8 mm thickness
- 4.5 mm in diameter

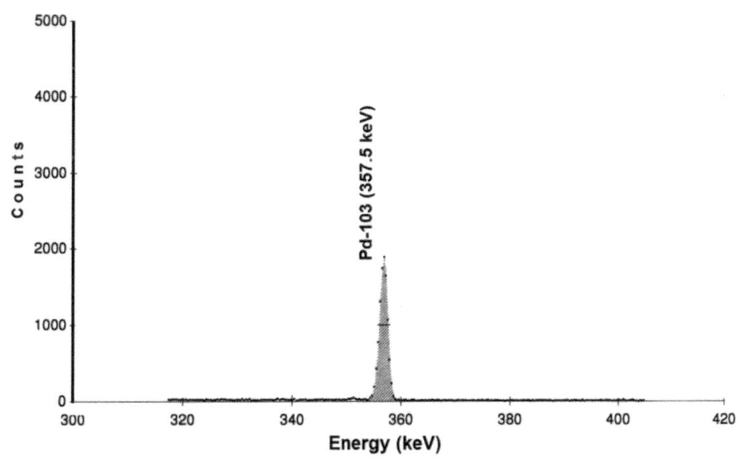

Figure 4.
HPGe spectrum of radiochemically separated ^{103}Pd [10].

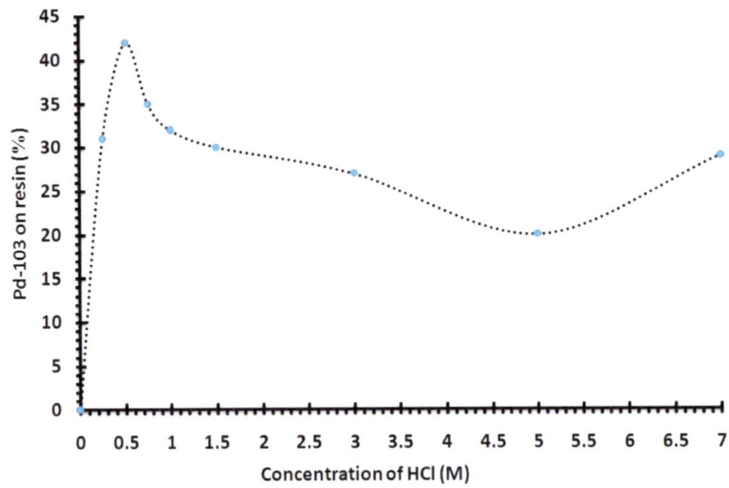

Figure 5.
Absorption profile of ^{103}Pd, as a function of HCl concentration on Amberlite®IR-93 resin [10].

For the TLD calibration, before each experimental measurement, the entire batch of TLDs is exposed to a calibrated Cobalt-60 standard beam. The variation of response of the TLDs to the same exposure is tracked by normalizing the individual TLD readings to the average value.

The irradiated TLDs (irradiated by the brachytherapy seed in the phantom) are read using a KFKI RMKI TLD reader (KFKI Research Institute of the Hungarian Academy of Sciences, Budapest, Hungary), and then they are annealed by heating at 240°C for 10 min followed by fast cooling. The responses of the TLD have to be corrected for background. This is done by subtracting the average response of background TLDs from the responses of all other TLDs in each measurement [25, 32].

3.3.2 Phantoms

To determine the dosimetric parameters of the seeds by experimental measurement, the phantom of Perspex slabs (**Figure 6**), by the following specification, has been used.

- Dimension: 30 cm × 30 cm × 15 cm
- Composition (by weight percent): H, 8%; C, 60%; and O, 32%
- Density: 1.19 g/cm^3

The design of the two phantoms to measure the radial dose and anisotropy functions are based on those of [25].

Figure 7 shows the phantom slab which is used for the experimental measurements of the brachytherapy seed radial dose function values. As shown in **Figure 7**, in the central phantom slaps, the holes are drilled to place the TLD circular chips. The circular surface of the TLDs is parallel to the seed long axis and is perpendicular to the slab plane [33, 34].

The measurements are carried out at distances of r = 0.5, 1, 1.5, 2, 3, 4, and 5 cm relative to the seed center. To minimize the interference of any of TLDs with regard to the response by other TLD chips, the measurement is performed in a spiral configuration [6, 8, 35, 36].

Figure 6.
Perspex phantom slabs.

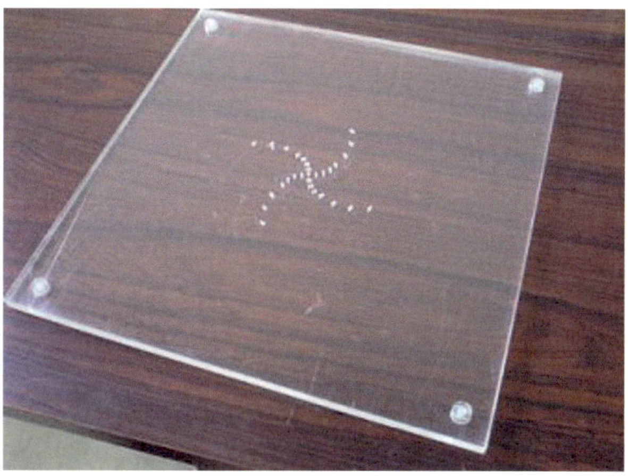

Figure 7.
Central slabs of Perspex phantoms used for the experimental determination of radial dose function values.

In this study, 28 TLDs (4 at each radial distance) were used for every single experiment to prevent the shadowing effect due to the configuration and the design of the phantom. To improve the statistical quality of the data, the experiment was repeated several times.

The other phantom is shown in **Figure 8**. This phantom has been used for the measurement of the anisotropy function of the brachytherapy seed. It has the same dimensions as the first phantom but differs in the configuration of the source in that. The source is placed parallel to the central slab plane with its long axis.

The TLDs are placed at radial distances of r = 1.5, 2, 3, and 5 cm relative to the seed center and also lie at the polar angles θ ranging from 0 to 330° in 30° increments with respect to the seed long axis. The measurements were performed with 48 holes containing TLDs since it was found that for the experimental anisotropy function determination at a specific point, shadowing effects due to the TLDs that lie at the same polar angle do not affect results. This is due to the definition of anisotropy

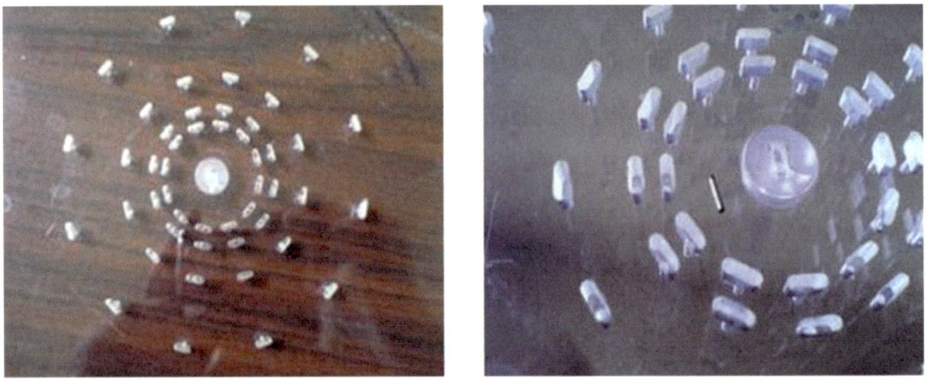

Figure 8.
Central slabs of Perspex phantoms used for the experimental determination of anisotropy function values.

function that normalizes dose rate at a particular (r, θ) point to the dose rate at the corresponding point along the transverse source bisector, (r, 90°). Therefore, since shadowing was found similar at any polar angle for the same radial distance, the overall effect is canceled out in the calculation of an anisotropy function [12].

4. Conclusion

This chapter presents the application of the cyclotron in brachytherapy by the production of radioisotopes such as Palladium-103.

In this chapter, production of the ^{103}Pd via cyclotron has been presented. **103Pd** is used in permanent low-dose radiation brachytherapy. So preparation of the brachytherapy source having ^{103}Pd radioisotope has also been discussed.

^{103}Pd production is performed via the ^{103}Rh(p,n)^{103}Pd reaction by 18 MeV protons for 15 h at 200 μA beam current. The optimum energy range and the thickness of the rhodium target are calculated by the several computer codes (ALIS/ASH, TALYS, EMPIRE). Several codes have been used to increase the accuracy of the calculations. To use the ^{103}Pd as brachytherapy source, the resin beads which are loaded by ^{103}Pd are encapsulated inside the titanium capsule, and then the capsules are implanted into the cancerous area. So, after the chemical separation process, ^{103}Pd radioisotope is absorbed uniformly in the resin Amberlite®IR-93, (20–50 mesh) bead to encapsulate them inside the titanium casing.

According to the American Association of Physicists in Medicine (AAPM) Radiation Therapy Committee recommendation, the dosimetric parameters of all new interstitial brachytherapy sources with energies less than 50 keV should be determined by two independent verifications, experimental measurements and theoretical calculations. The method for the theoretical calculation of the brachytherapy seed has been previously explained in Refs. [6, 9]. In this study, the experimental measurement method, the design, and dimension of the phantom and configuration of the TLDs have also been explained.

Author details

Pooneh Saidi[1*] and Mahdi Sadeghi[2]

1 Parsikan Iran Engineering and Management Consultants, Tehran, Iran

2 Medical Physics Department, School of Medicine, Iran University of Medical Sciences, Tehran, Iran

*Address all correspondence to: poonehsaidi@gmail.com

References

[1] Khan F. Handbook of the Physics of Radiation Therapy. Baltimore: Williams and Wilkins; 2009. pp. 36-53

[2] Schlyer DJ, Van den Winkel P, Ruth TJ, Vora MM. Cyclotron produced radionuclides: Principles and practice. In: Technical Reports Series No. 465. Vienna: IAEA; 2008. pp. 31-57

[3] Simone Manenti S, del Carmen M, Santoro A, Cotogno G, Duchemin C, Haddad F, et al. Excitation function and yield for the 103Rh(d,2n)103Pd nuclear reaction: Optimization of the production of palladium-103. Nuclear Medicine and Biology. 2017;**49**:30-37

[4] Baltas D, Sakelliou L, Zamboglou N. The Physics of Modern Brachytherapy for Oncology. New York: Taylor & Francis Group; 2007. p. 165

[5] Sadeghi M, Enferadi M, Shirazi A. External and internal radiation therapy: Past and future directions. Journal of Cancer Research and Therapeutics. 2010;**6**:239-248

[6] Saidi P, Sadeghi M, Tenreiro C. Theory and applications of Monte Carlo simulations. In: Chan VWK, editor. Variance Reduction of Monte Carlo Simulation in Nuclear Engineering Field, Ch. 7. London: Intech; 2013. pp. 153-172. DOI: 10.5772/53384

[7] Saidi P, Sadeghi M, Shirazi A, Tenreiro C. Monte Carlo calculation of dosimetry parameters for the IR08-103Pd brachytherapy source. Medical Physics. 2010;**37**:2509-2515. DOI: 10.1118/1.3416922

[8] Saidi P, Sadeghi M, Tenreiro C. Experimental measurements and Monte Carlo calculations for 103Pd dosimetry of the 12 mm COMS eye plaque. Physica Medica. 2013;**29**:286-294. DOI: 10.1016/j.ejmp.2012.04.002

[9] Sadeghi M, Saidi P, Tenreiro C. Dosimetric characteristics of the brachytherapy sources based on Monte Carlo method. In: Mode CJ, editor. Applications of Monte Carlo Methods in Biology, Medicine and Other Fields of Science, Ch. 10. London: Intech; 2011. pp. 155-176. DOI: 10.5772/15884

[10] Saidi P, Sadeghi M, Enferadi M, Aslani G. Investigation of palladium-103 production and IR07-103Pd brachytherapy seed preparation. Annals of Nuclear Energy. 2011;**38**:2168-2173

[11] Sadeghi M, Shirazi B. Extraction separation of no carrier- added ^{103}Pd from irradiated Rh target, Cu and Zn using α-furyldioxime, dimethylglyoxime and α-benzildioxime. Applied Radiation and Isotopes. 2008;**66**:1810-1813

[12] Sadeghi M, Shirazi B, Shadanpour N. Solvent extraction of no-carrier-added 103-Pd from irradiated rhodium target with α-furyldioxime. Journal of Radioanalytical and Nuclear Chemistry. 2006;**269**:223-225

[13] Broeders CHM, Konobeyev A, Kocrovin YA, Blann VPM. Pre-compound and Evaporation Model Code System for Calculation of Excitation Functions, Energy and Angular Distributions of Emitted Particles in Nuclear Reactions at Intermediate Energies. Frschungszentrum Karlsruhe Report FZKA. Hannover; 2006. p. 7183

[14] Koning AJ, Hilaire S, Duijvestijn M. TALYS-1.2: A Nuclear Reaction Program, User Manual. Netherlands: NRG; 2009

[15] EXFOR/CSISRS. 2011. Experimental Nuclear Reaction Data. Available from: http://wwwnds.iaea.org/exfor

[16] Koning AJ, Rochman D. TENDL-2010: TALYS-Based Evaluated Nuclear

Data Library. The Netherlands: Nuclear Research and Consultancy Group (NRG) Petten; 2010. Available from: http://www.talys.eu/tendl-2010

[17] Saidi Bidokhti P, Sadeghi M, Fateh B, Matloobi M, Aslani G. Nuclear data measurement of 186-Re production via various reactions. Nuclear Engineering and Technology. 2010;**42**(5):600-607

[18] Sadeghi M, Enferadi M. Nuclear model calculations on the production of 119Sb via various nuclear reactions. Annals of Nuclear Energy. 2011;**38**:825-834

[19] Büyükuslu H, Kaplan A, Tel E, Aydin A, Yildırım G, Bolukdemir MH. Theoretical cross-sections of ^{209}Bi, ^{232}Th, ^{235}U, and 238U on deuteron-induced reactions. Annals of Nuclear Energy. 2010;**37**:534-539

[20] Herman M, Capote R, Zerkin V, Trkov A, Wienke H, Sin M, et al. EMPIRE Modular System for Nuclear Reaction Calculations (version: 3.1 Rivoli). 2011. Available from: https://ndclx4.bnl.gov/gf/project/empire/

[21] Ziegler JF, Biersack JP, Littmark U. The Code of SRIM, the Stopping and Range of Ion in Matter. New York, USA: IBM-Research; 2006

[22] Saidi P, Sadeghi M. Theory, application and implementation of Monte Carlo method in science and technology. In: Bidokhti PS, editor. Modeling, Simulation, and Dosimetry of 103-Pd Eye Plaque Brachytherapy, Ch. 2. London: Intech; 2019. pp. 19-38. DOI: 10.5772/intechopen.78141

[23] Raisali G, Ghonchehnazi MG, Shokrani P, Sadeghi M. Determination of the geometry function for a brachytherapy seed, comparing MCNP results with TG-43U1 analytical approximations. Nukleonika. 2008;**53**:45-49

[24] Ataeinia V, Raisali G, Sadeghi M. Determination of dosimetry parameters of ADVANTAGETM 103Pd brachytherapy seed using MCNP4C computer code. Nukleonika. 2009;**54**:181-187

[25] Saidi P, Sadeghi M, Hosseini SH. Thermoluminescent dosimetry of the IR06-103Pd brachytherapy source. Journal of Applied Clinical Medical Physics. 2011;**12**(4):286-295. DOI: 10.1120/jacmp.v12i4.3581

[26] Harper PV, Lathrop K, Need JL. The thick target yield and excitation function for the reaction 103Rh(p,n)103Pd. Nuclear Science. 1961;**15**:124

[27] Hansen LF, Albert RD. Statistical theory predictions for 5 to 11 MeV (p,n) and (p,p') nuclear reactions in 51V, 59Co, 63Cu, 65Cu, and 103Rh. Physics Review. 1962;**128**:291-299

[28] Hermanne A, Sonck M, Fenyvesi A, Daraban L. Study of production of ^{103}Pd and characterisation of possible contaminants in the proton irradiation of 103Rh up to 28 MeV. Nuclear Instruments & Methods. 2000;**B170**:281-292

[29] Sudar S, Cserpak F, Qaim SM. Measurements and nuclear model calculations on proton-induced reactions on Rh-103 up to 40-MeV-evaluation of the excitation function of the Rh-103(p,n)Pd-103 reaction relevant to the production of the therapeutic radionuclide Pd-103. Applied Radiation and Isotopes. 2002;**56**:821-831

[30] Tárkányi F, Hermanne A, Király B, Takacs S, Ditrói F, Csikai J, et al. New cross-sections for production of 103Pd; review of charged particle production routes. Applied Radiation and Isotopes. 2009;**67**:1574-1581

[31] Hermanne A, Sonck M, Takacs S, Tárkányi F, Shubin FY. Study on alternative production of 103Pd and characterisation of contaminant sin the deuteron irradiation of 103Rh up to 21 MeV. Nuclear Instruments & Methods. 2002;**B187**:3-14

[32] Duggan L, Hood C, Warren-Forward H, Haque M, Kron T. Variation in dose response with X-ray energy of LiF:Mg, Cu, P thermoluminescence dosimeters: Implications for clinical dosimetry. Physics in Medicine and Biology. 2004;**49**(18):3831-3845

[33] Meigooni AS, Gearheart DM, Sowards K. Experimental determination of dosimetric characteristics of bests 125I brachytherapy source. Medical Physics. 2000;**27**(9):2168-2173

[34] Meigooni AS, Sowards K, Soldano M. Dosimetric characteristics of the intersource 103-palladium brachytherapy source. Medical Physics. 2000;**27**(5):1093-1099

[35] Hosseini SH, Sadeghi M, Ataeinia V. Dosimetric comparison of four new designs 103-Pd brachytherapy sources: Optimal design using silver and copper rod cores. Medical Physics. 2009;**36**(7):3080-3085

[36] Sadeghi M, Hosseini SH, Raisali G. Experimental measurements and Monte Carlo calculations of dosimetric parameters of the IRA1-103Pd brachytherapy source. Applied Radiation and Isotopes. 2008;**66**(10):1431-1437

Chapter 8

Bisphosphonates as Chelating Agents in Uranium Poisoning

Adriana Beatriz Martínez, Carola Bettina Bozal, Nadia Soledad Orona, Deborah Ruth Tasat and Angela Matilde Ubios

Abstract

The study of uranium toxicity is very important for public health in general and especially for workers involved in the processes of uranium mining and milling because of the immediate and/or mediate risks of exposure. Most available studies show unsuccessful attempts to eliminate uranium from target organs once the poisoning has occurred. Our group has managed to avoid damage to target organs (short-term kidney and long-term bone damage) in a high percentage of animals treated with lethal doses of uranyl nitrate through the effective chelating action of a single dose of bisodic etidronate. In this context, the contributions of our team and other groups working on chelating therapies provide a starting point for progress in the search for agents for preventing and/or reducing the toxic effects of uranium.

Keywords: uranium, poisoning, bisphosphonates, bisodic etidronate, chelating agents

1. Introduction

As from the World War II, increasing interest in nuclear power brought about an increase in uranium exploration and the development of new plants for uranium processing and manufacture in many countries. These activities may involve accidental occupational poisoning for workers, and no protocol has yet been designed for rapid application in these cases to prevent rapid-onset life-threatening complications due to kidney failure. As the uranium-related industry increases, so does the potential for these accidental poisonings. In this chapter we present and discuss the papers published by our laboratory over almost 40 years of research in the field of the toxicology of uranium in animal models under the direction of Dr. Romulo Luis Cabrini, which replicated hypothetical situations of acute accidental exposure to uranium via different routes of entry. Our group has specifically studied toxicity on target organs (kidneys and bone) and developed a potentially effective protocol with bisodic etidronate used as a chelating agent that prevents kidney failure and bone alterations. We are currently processing histological and histomorphometrical kidney and bone samples from a long-term experiment with animals poisoned with lethal doses of uranyl nitrate and treated with a single subcutaneous dose of bisodic etidronate which survived for 1 year after treatment with good quality of life, similar to that of controls.

IntechOpen

2. Biological impact of uranium: pharmacokinetics

Because uranium (U) is present in food, air, soil, and water, humans are constantly exposed to certain amounts of this element. Notwithstanding, the biological impact of such natural exposure on human physiology and pathophysiology is not yet fully known. However, it is known that overexposure to U may result in toxicity, which is derived from an excessive accumulation of the element in the organism. This accumulation depends on various factors, including route of entry, duration of exposure, dose, chemical compound of which it forms part, and absorption [1, 2].

Uranium can enter the body through different routes: oral, inhalation, percutaneous, or subcutaneous. Regardless of the route of entry, absorbed U enters systemic circulation, is distributed in the organism, and accumulates mainly in the bones (66%), kidneys (8%), and liver (16%) [3]. Approximately 1–5% of an oral dose is absorbed in the digestive tract [4], and nearly 60% of U is eliminated rapidly from the blood and slowly from organ depots with the urine by renal mechanisms in the first 24 h [5, 6]. In rats, most of the absorbed U is eliminated by the kidneys in a few days; half of it is excreted within 2–6 days [7] and the rest within 7 days. Ninety-five percent (95%) of U present in the kidneys of intoxicated rats is excreted in the urine within a week, and very small amounts remain in other organs [6, 8]. Uranium compounds can dissociate and form new compounds with various organic and inorganic anions. In body fluids, tetravalent uranium (+4) tends to oxidize to the hexavalent form (+6) followed by uranyl ion formation. Experimentally, using animal models, it was shown that uranyl ions are associated with ultrafiltered low molecular weight serum proteins, transferrin, and other plasma proteins [3]. In 2005, Vidaud *et al.* [9] were able to identify uranium-binding proteins in human serum fractions by means of an in vitro-sensitive procedure involving a combination of bidimensional chromatography with time-resolved fluorescence, coupled with proteomic analysis. These authors demonstrated that not all targets are metalloproteins, suggesting that uranyl ions can use a wide variety of binding sites, thus providing additional insights for a better understanding of uranium chemical toxicity. U also binds to phospholipids and membrane proteins of proximal contoured tubules [10].

On the other hand, when injected intravenously, almost 50% of the U is eliminated, while the other 50% is deposited in the skeleton (25%) and in soft tissues (25%), mainly kidneys. The U deposited in extrarenal soft tissues—mainly liver and spleen—is removed very slowly [11]. Orcutt *et al.* [12] reported that the percutaneous route constitutes an effective route of entry for soluble U compounds. De Rey *et al.* [13] demonstrated that uranyl nitrate (UN) can penetrate through the skin of adult Wistar rats in approximately 15 min and accumulate in the intercellular spaces between the granular and corneal layers. After 48 h, the U was not found in the skin, and the animals experienced signs of severe toxicity ranging from weight loss to death, clearly indicating its passage from this organ into the circulation. The retention of U particles after inhalation depends on the size of the particulate and the type of U compound. Harris *et al.* [14] reported that insoluble compounds (uranium dioxide and uranium trioxide) with average particle size below 2 μm have very long biological half-lives (120 days or more).

Uranium impact on human health may come from abandoned hard rock mines, which can contaminate the three natural resources—water, air, and soil—thus becoming a potential source of chronic exposure and toxicity for individuals living in the area.

It is worth noting that U toxicity depends on several factors, such as sex, age, body mass index [15], and species. Of all the mammals studied, humans seem to be the least sensitive to U [16]. Still, overexposure to U may cause pathological alterations in the different organs in both humans and animals.

2.1 Oral route

The oral route is important because of the possibility of the general population having to ingest U-contaminated water or food continuously, as well as the risk of workers ingesting toxic and/or lethal doses during accidents in some of the steps of the enrichment process. The literature contains interesting—though limited—data on the incorporation of U via the oral route, obtained from studies in both experimental animals and humans.

Harrison *et al.* [17] studied the gastrointestinal absorption of two U compounds administered orally by gastric intubation (gavage) in hamsters and demonstrated that soluble UN absorption was seven times greater than insoluble uranium dioxide (UD) absorption.

La Touche *et al.* [6] investigated the absorption and kinetics of UN administered orally in a model with fasting adult male Wistar rats that replicate the human intake of contaminated water after a night of fasting. It produced more absorption of the compound than if the rats had not been subjected to fasting.

Anke *et al.* [18] studied wild plants and cultivated plants from the immediate vicinity of uranium waste dumps and found that they stored up to eightfold higher uranium concentrations than controls.

In 2013, the ATSDR (US Agency for Toxic Substances and Disease Registry) [1] established a minimum risk level for the ingestion of U, known by the acronym MRL (minimum risk level), which is 0.002 μg of U per kilo of weight per day. The MRL is an estimate of the daily exposure of a human being to a dangerous substance that probably does not represent an appreciable risk of adverse effects (excluding cancer) beyond the time of exposure. MRL values for both oral and inhalation routes vary with different exposure times: acute (1–14 days), intermediate (15–365 days), and chronic (365 days or longer).

Zamora *et al.* [19] studied the effects of chronic ingestion of U in humans after drinking contaminated water. They suggested that intake of U doses such as those found in some underground water wells over prolonged periods of time altered the renal function. These observed effects may represent a manifestation of subclinical toxicity that does not necessarily lead to renal dysfunction or obvious injury. Instead, it may be the first stage in a spectrum in which chronic intake of elevated U levels can lead to irreversible renal injury [20].

In addition to being ingested via drinking water, U can be taken in orally through contaminated food, mainly beef and fish.

The concentrations of U in fish muscle (per gram of dry weight) extracted from a Canadian lake contaminated by tributaries of a U processing plant were 7–11 times higher than in those from uncontaminated lakes [21]. Lapham *et al.* [22] analyzed U concentrations in cattle muscle and found that although U levels in the muscle of the exposed cattle were almost imperceptible with respect to the controls, U concentration in the liver and renal tissues was 4 times higher in the exposed cattle than in the control. Moreover, U levels in the bone (femur samples) were found to be 13 times higher than in the controls.

Other than beef and fish, some underground vegetables such as potatoes, sweet potatoes, and turnips contribute approximately 38% of the U intake per diet according to the general food consumption rate [23].

Although there are no studies in humans regarding the lethal effects of U and its compounds when ingested orally, several animal studies have shown that a very high intake dose of U can be lethal for acute (1–14 days), subacute (14–365 days), and chronic (more than 365 days) exposures [24].

Maynard and Hodge [1] obtained an LD_{50} (lethal dose 50 population %) value for oral (UN) of 1579 mg of U/kg/day in rats of both sexes (without specifying strain) in a 30-day study. Maynard *et al.* [25] found 16% mortality in rats that were intoxicated with 664 mg/kg/day of U for 30 days orally. In our laboratory, Martinez *et al.* [26] studied the effects of a lethal dose of UN administered by gavage in male mice: we found that 350 mg/kg was the LD_{99} (lethal dose 99 population %) on the third day of the experiment.

Studies in rats suggest that the primary pathway for gastrointestinal absorption of soluble U is through the small intestinal epithelium [27, 28] via the transcellular pathway [29]. In the event of ingestion, the digestive tract is the first biological system exposed to U intake via the intestinal lumen. However, little research has addressed the biological consequences of contamination with U on intestinal properties such as the barrier function and/or the immune status of this tissue. Dublineau *et al.* [28, 29] studied both acute contamination with Depleted Uranium (DU) at high doses and chronic contamination at low doses on inflammatory reactions in the intestine when orally delivered. They found that acute and chronic ingestion of DU modulated expression and/or production of cytokines in the intestine and had similar effects to those observed with lead on the nitric oxide pathway.

2.2 Inhalation route

Inhalation is a major route of human exposure to environmental particles. When inhaled, U particles, based on their size, may be deposited on the lung ciliated epithelial lining or may reach the lower respiratory tract. Small particles of U containing dust could be inhaled by U miners and people living close to the mines, penetrating deeply into their lungs. The particles can be phagocytosed by alveolar macrophages (AM) and/or cross the alveolar capillary barrier, thereby reaching the bloodstream.

Uranium toxicity depends, among other things, on the solubility, dose, and route of exposure. In general, the more insoluble U compounds (uranium trioxide, uranium dioxide, uranium peroxide, and triuranium octaoxide) have greater potential for long-term effects in the lung, probably due to the long-term retention of the compound [1]. On the contrary, soluble U compounds (uranyl fluoride, 1 uranium tetrachloride, and uranyl nitrate hexahydrate), due to their easier absorption in the lungs and passage into the bloodstream, exert their action mainly on the extrapulmonary organs [1]. Thus, lungs and extrapulmonary organs are susceptible to negative impact by U particles.

Even though U can enter the body through different routes, macrophages are always the first cells to respond to these or any other xenobiotic agents. Primary cultured AM is a suitable *in vitro* model for studying the effects of U at cell level and its cytotoxic mechanism. Tasat and de Rey [30] demonstrated the adverse effects of insoluble uranium dioxide using AM obtained from rat bronchoalveolar lavage. The study revealed the ability of macrophages to phagocyte U particles in a short time despite the high toxicity that metal exerts on cell membranes. The ultrastructural analysis detected the U particles confined within intracytoplasmic vacuoles or free in the macrophage cytoplasm, which in turn could lead to cell death. More recently, Orona *et al.* [31] demonstrated that exposure to the soluble U compound UN, also induced cell death in cultured rat AM. The cytotoxic mechanism studied in this *in vitro* model showed that at low doses, UN stimulated phagocytosis and generation of superoxide anion (O_2^-), while at high doses, it induced the secretion of TNFα. Therefore, Orona *et al.* suggested that cell death at low doses was principally mediated by reactive oxygen species (ROS), while the signaling pathway when exposed to higher UN doses was principally mediated by the release of pro-inflammatory mediators inhibiting the generation of ROS.

These authors hypothesized that an oxidant-antioxidant imbalance provoked by activated macrophages after U inhalation may lead to an alteration in macrophage metabolism. This cell response may in turn modify the pulmonary tissue microenvironment, thereby participating in the development of lung pathologies associated with U exposure. Uranium exposure involves both an occupational risk for miners and workers who handle it constantly, and an environmental hazard to the health of the population at large. Katz *et al.* [32] updated the chemistry, pharmacokinetics, and toxicological effects of U on several systems in the mammalian body which were previously reviewed by Craft *et al.* [33] and Briner [34]. However, the adverse impact on human health is still controversial. In this context, we studied the effect of UN on macrophages differentiated from human THP-1 monocytes. As is clearly shown in **Figure 1**, increasing doses of soluble UN provoked a significant decrease in macrophage-like cell viability, similar to what was observed with murine macrophages. Extrapolated to the *in vivo* situation, these findings might help to explain, in part, how acute or chronic inflammatory states observed in U-exposed individuals with DNA lesions could be linked to the U compounds dose.

Despite all the research conducted by the scientific community to clarify and understand the effect of both insoluble and soluble U compounds, the data reported appear to have somehow been compromised by the political agendas of special interest groups at both national and international levels.

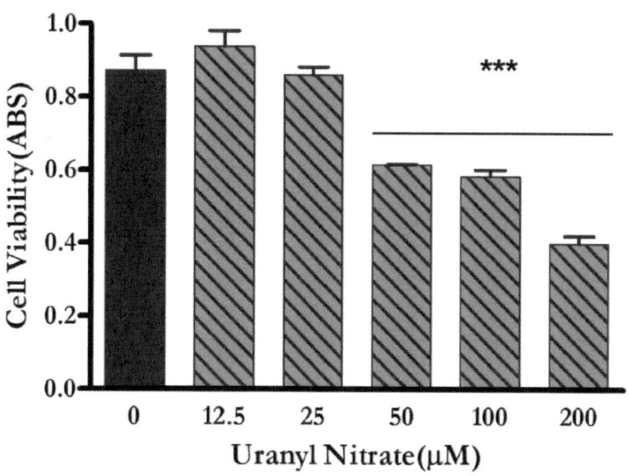

Figure 1.
Cell viability of macrophages differentiated from human THP-1 monocytes exposed to UN. Higher doses of UN caused a reduction in cell viability. Values are represented as mean ± SD. Results are compared employing one-way ANOVA followed by Dunnett's test P < 0.001.

2.3 Dermal contact

2.3.1 Percutaneous absorption

Chemically induced renal failure caused 100% mortality in male Wistar rats after five daily exposures to 237 or 1928 mg U/kg/day as UN hexahydrate or ammonium uranyl tricarbonate, respectively, applied in a water-Vaseline® emulsion. A 60% mortality rate was also reported for other male Wistar rats that received daily applications of 1965 mg U/kg as uranyl acetate dihydrate for 1–11 days. No death was reported for other Wistar rats similarly treated with 2103 mg U/kg/day as ammonium diuranate or to an unspecified dose of UD [13].

Decreased survival was observed in female Wistar rats following dermal application of 280 mg U as UN hexahydrate diluted in an oil–water emulsion; survival was inversely related to the duration of exposure and the application area [35]. A

24-h application to areas of 0.5, 1, 2, 4, 6, 8, or 16 cm^2 resulted in survival rates of 80, 83, 67, 29, 33, 0, and 0%, respectively; application to 8 cm^2 for 1 min, 7 min, 15 min, 30 min, 1 h, 8 h, or 24 h resulted in survival rates of 100, 100, 100, 67, 45, 43, 10, and 0%, respectively.

2.3.2 Subcutaneous absorption

Subcutaneous or intradermal U contamination takes place in the presence of a wound. This poses a real risk to workers handling U dust on a daily basis and to soldiers who fought in the modern wars (Balkan, Gulf, etc.). Penetration of DU shrapnel bullets into the skin has become the focus of increasing attention. In fact, the only documented cases of exposure to U are those of the Gulf War veterans who retained DU shrapnel fragments [36].

Subcutaneous implantation of insoluble UO_2 was investigated in our laboratory in an experimental animal model in rats by de Rey *et al.* [37]. This group showed that animals receiving doses higher than 0.01 g/kg died within the first 6 days due to acute renal failure. Histological analysis revealed the presence of deposits of uranium taken up by macrophages at 24 and 48 h postexposure. Deposits were found between the endothelial cells and the renal parenchyma, suggesting that the U insoluble compound implanted subcutaneously is transported and deposited.

3. Uranium toxicity and its main target organs

Regardless of the route of entry, U has two main target organs: the kidney and the bone. The magnitude of the adverse U impact in these two organs is dose- and time-dependent.

3.1 Renal toxicity

In 1991, the World Health Organization (WHO) identified U and other heavy metals such as lead, mercury, and cadmium as nephrotoxic elements. Early in 1949, Voegtlin and Hodge [38], in the framework of the Manhattan Project, studied the toxic effects of U in animal experimental models, finding characteristic histological features of kidney injury, regardless of the route of entry to the organism. Subsequently, many animal studies have shown that inhalation, oral exposure, or dermal exposure to uranium results in kidney damage [35, 39–41]. This damage was histologically manifested principally as glomerular and tubular wall degeneration. After being filtered by the renal glomerulus, the uranium-bicarbonate complex enters the glomerular urine. The bicarbonate is reabsorbed into the venous blood and loses the uranyl ion at the time of its passage through the proximal contoured tubule. The uranyl ion reacts with the membrane proteins of the cylindrical cells causing cell damage, and, when doses are high, cell death can occur, releasing the cell content into the urine [42]. Ultrastructural analysis showed damage to the endothelial cells in the glomerulus, such as loss of cell processes and reduction in the density of the endothelial fenestrae [43–46]. Although the most obvious effect of U exposure is damage to the proximal convoluted tubules, necrotic cells from the tubular epithelium have also been reported [19]. The histological alterations observed as a result of exposure to UN include partial degeneration, necrosis, and cast formation in proximal convoluted tubule although with damage to brush border, but glomeruli remain intact [47]. In our laboratory, we have observed that kidneys of exposed animals revealed the usual U-induced tubule necrosis lesions with abundant hyaline cylinders and extensive areas of necrosis after 48 h of 350 mg UN/kg b.w. exposure.

At glomerular level, although glomerulus structure appears to be intact, widening of the Bowman's capsule is evident [41] (**Figure 2**).

The analysis of autoradiographic, histological, and renal functioning studies showed that the site where the U produces greatest injury is the distal third of the proximal contoured tubule. If the tubular cells are not too damaged, they can repair the alterations and regenerate. Recovery is rapid despite the fact that the regenerated cells are atypical in some details, and within a few weeks or a month, both the biochemical parameters and renal histology are normalized. In 1982, Haley [44] studied different U compounds with special interest in their effects on the renal parenchyma. Of all the compounds studied, UN proved to be the most nephrotoxic compound, which explains why it is frequently used to produce experimental renal failure.

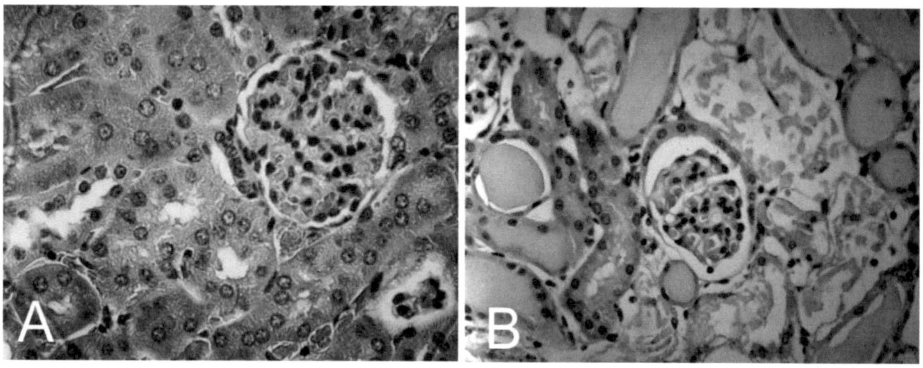

Figure 2.
Histopathology of kidney tissue from control and UN exposure groups. Cortical region of kidney: (A) control animal, intact tubular epithelium with no damage and (B) animal exposed to 350 mg UN/kg b.w., 48 h, tubular damage, loss of microvilli, extensive necrosis and cast formation in tubules, and necrosis of proximal tubular epithelium. Marked vacuolization of the tubules, abundant hyaline cylinders, and extensive areas of necrosis. Widening of the uriniferous tubule and of Bowman's capsule is also evident (HE 400X).

3.2 Effects on bone tissue

The incorporation of U compounds into bone tissue has been demonstrated by biochemical analysis, autoradiographic methods, neutron activation analysis, and X-ray microanalysis. Many U isotopes are considered more as a chemical risk than as a radiological risk. The radioactivity of UN can be considered negligible, since the radioactivity of ^{238}U, as we have seen, is very low [48–50]. This would explain that bone formation alterations could be due preferably to the chemical toxicity of U [51]. As mentioned above, Hursh *et al.* [11] reported that 25% of the systemically administered U is deposited in the skeleton and tends to bind to the newly formed bone. Since the bone is the only tissue in which U deposits can be found a long time after exposure, it is considered to be the critical organ in chronic exposures, displacing the kidney as a target. Autoradiographic studies demonstrated initial U deposits on the surfaces of the endosteum, the periosteum, and the haversian bone, particularly in areas where calcification occurs. Back in 1948, Neuman *et al.* showed that there was an association between U deposits and bone formation [52], but at that time, it was not known whether this process was affected. Three decades later, for the first time, it was found that U also affects bone metabolism in acute poisoning. Guglielmotti *et al.* [53, 54] were the first to demonstrate, by means of histological and histomorphometric methods, that the inhibition of bone formation was a result of acute intoxication with UN. The study was carried out with Wistar rats (weighing 90 g/b.w.) intoxicated with a dose of 2 mg/kg of UN applied intraperitoneally, and observations were made at the level of the endochondral ossification of the tibia. After 14 days, inhibition of endochondral ossification of the tibia was observed in the intoxicated animals but not in the control animals. In a histomorphometric analysis of the bone of intoxicated animals, they found a decrease in bone surfaces covered by active osteoblasts and the consequent increase in inactive osteoblasts. Such results were attributed to U, which has been suggested to cause alterations in the osteoblast differentiation process or in their cell precursors. At the same time, the remaining osteoblasts may form the sealing trabeculae that were seen in the metaphyseal bone.

In 1987, Guglielmotti *et al.* [55] used the same animal model to study the effect of a low dose of UN (0.8 mg/kg) on tooth extraction socket healing over time (14, 30, and 60 days post-surgery). Results revealed a delay in socket healing with respect to controls. At low doses, U exerted its toxic effect on the recruitment and/or differentiation of osteoblasts, despite the cell damage, and dramatically inhibited bone formation observed after acute poisoning.

Ubios *et al.* [56] confirmed the results described above through ultrastructural studies in which they also detected the presence of small electron-dense deposits in the osteoblast cell membrane, inferring that they were U particles. The inhibitory UN effect on bone formation was evidenced as a reduction in bone growth of tibiae and mandibles [57, 58] and as a delay in tooth eruption [59]. In 2007, Tasat *et al.* [60] showed that UN modified osteoblast cell metabolism by increasing reactive oxygen species generation and reducing alkaline phosphatase activity, suggesting that ROS could play a more complex role in cell physiology than simply by causing oxidative damage.

The effects of U on bone formation are evident not only in situations of active ossification, but we have also observed them in bone modeling and remodeling, where, in addition, the reduction in bone formation activity was associated with an increase in bone resorption [61]. Nevertheless, while many studies have focused on the effect of U on bone formation and osteoblasts, the impact of U on bone resorption has been poorly explored. In this context, Ubios *et al.* [61] conducted a pioneering study demonstrating an increase in resorption of the alveolar bone of the jaws after intraperitoneal injection of UN in Wistar rats. Subsequently, in our laboratory, we have also observed a histomorphometrical increase in resorption surfaces in metaphyseal bone after oral administration of a lethal oral dose of UN [62]. Recent studies have revealed dose- and time-dependent U cytotoxicity on pre-osteoclast cell lines, impairing osteoclast formation and function [63].

4. Chelating agents in uranium poisoning

Uranium toxicity has been a concern for over 100 years. The toxicology of many forms of uranium, ranging from dust of several oxides to soluble uranyl ions, was thoroughly studied during the Manhattan Project in the United States in the 1940s [64]. Data available in the literature show that most studies have focused on finding a compound that accelerates U decorporation after it has reached the target organs: kidney (acute intoxication) and bone (chronic exposure).

From a pharmacological standpoint, different methods have been tested to counteract the toxic effect of U. Several chelating agents such as EDTA, Tiron, DTPA, or aminosalicylic acid have been experimentally assayed. However, even when these agents are able to reduce mortality, none of them achieve 100% survival. Bicarbonate can be administered to reduce U body burdens due to acute exposures. Bicarbonate ions form a complex with U and alkalize the blood, both of which enhance the excretion from the kidneys by glomerular filtration [65], and such an application was described in a case of prophylactic treatment [66]. Experimental evidence in animals indicates that chelation therapy may reduce the body burden of U. Several compounds were found to enhance the urinary and fecal excretion of U if administered soon after U exposure. When administered immediately after exposure to U, Tiron® (sodium 4,5-dihydroxybenzene-1,3-disulfonate) resulted in the greatest reduction in renal and bone levels of U and acute lethal effects in animals [67, 68]. None of the chelating agents affected bone levels of U when administered ≥24 h after exposure to U [68]. Bicarbonate treatment is also limited to very near-term exposures. Another study that tested Tiron alone and in conjunction with either DTPA or ethylenediamine-N,N=-bis(2-hydroxyphenylacetic acid) (EDHPA) found that it reduced the U body burden by no more than about 35%, indicating that the administration of Tiron® is of limited practical value for the treatment of U exposures that do not greatly exceed the permitted intake level [69]. Our group began working with a bisphosphonate, ethane-1-hydroxy-1,1-bisphosphonate (EHBP) in 1986 based on the beneficial effect of bisphosphonates on bone when they are used in correct doses. Ubios *et al.* [70] showed the attenuation of the inhibitory effect of radiation on bone formation when the animals were treated with ethane-1-hydroxy-1,1-bisphosphonate (EHBP). In acute intoxication, U not only inhibits bone formation, but its excretion in urine also causes renal damage. The former effect is ameliorated by tetracycline (TC), probably due to its chelation property, which might also prevent U deposition in bone. Chemical determination of U incorporated in the bone and a histological study of the kidneys were performed by Guglielmotti *et al.* [71] on animals injected with U and then treated with TC. The results showed that TC was unable to prevent the binding of U to the bone, while it exacerbated U-induced renal damage. Ubios *et al.* [72] reported the beneficial effect of ethane 1-hydroxy-1,1-diphosphonate (EHDP) in restoring the inhibition of bone formation in cases of acute U intoxication in a post-extraction wound healing rat model. Ubios *et al.* [57] showed the use of a single subcutaneous injection of ethane-1-hydroxy-1,1-bisphosphonate to prevent mortality due to uranium poisoning in a rat model.

Of the chelation therapies that have been studied in animals, it appears that citric acid and citrate salts might be the most practical to employ in conflict areas where DU weapons have been used. Citric acid/citrate consumption should be recommended to anyone in areas where uranium aerosols might be found. Fructose and/or sucrose that may be present in some beverages that contain high levels of citrates should be avoided [64].

4.1 Bisphosphonates as chelating agents to avoid lethal poisoning by uranium

Bisphosphonates are used clinically to prevent osteoclastic bone resorption. Their use tends to achieve positive bone formation/resorption balance. These compounds are used to treat osteoporosis, a very common lesion in postmenopausal women in which the bone formation/resorption equation is altered (predominantly resorption), and the final matrix tends to decrease in volume, sometimes reaching mechanically dangerous situations due to fractures, among other problems.

The biological effects of bisphosphonates (BPs) as inhibitors of calcification and bone resorption were first described in the late 1960s. In the 50 years that have elapsed since then, BPs have become the leading drugs for the treatment of skeletal disorders characterized by increased bone resorption, including Paget's disease of the bone, bone metastases, multiple myeloma, osteoporosis, and childhood inherited bone disorders as osteogenesis imperfecta. The discovery and development of BPs as a major class of drugs for the treatment of bone diseases is a paradigm for the successful journey from "bench to bedside and back again." Several of the leading BPs achieved "blockbuster" status as branded drugs. However, these BPs have now come to the end of their patent life, making them highly affordable. The opportunity for new clinical applications for BPs also exists in other areas of medicine such as aging, cardiovascular disease, and radiation protection.

In our laboratory, based in data published by Ubios *et al.* [57], we designed animal experiments to focus on the potential of EHBP to prevent death after the administration of lethal oral doses of UN. In 2000, Martinez *et al.* [26] demonstrated, for the first time, that a single administration of EHBP is effective in reducing the lethal effect of U, and it is at least as useful as subcutaneous administration for prompt therapy of oral U exposure, achieving a survival rate of almost 50%. Tubule necrosis lesions were present in kidneys of mice intoxicated with UN, whereas lesions were less severe in mice treated with EHBP.

Based on the aforementioned results, Martinez *et al.* [41] evaluated the efficacy of EHBP in preventing renal dysfunction induced by a lethal dose of UN, employing serum levels of urea and creatinine as endpoints. Two experiments were performed with different time periods: 48 h and 14 days in male Balb/c mice with 25 g average body weight. Three of these groups received 350 mg/kg body weight of UN by gavage (forced oral administration). Two of the three exposed groups were treated with EHBP either by gavage in a dose of 500 mg/kg body weight or with a subcutaneous injection of 50 mg/kg body weight. The fourth group served as control. Urea and creatinine serum levels were markedly lower at 48 h in exposed animals treated with EHBP than in untreated exposed animals. On day 14 these values in exposed and treated animals did not differ significantly from control values. The renal function of animals treated with oral or subcutaneous EHBP that survived UN exposure was markedly improved compared to the controls of untreated exposed animals at 48 h. At 14 days, treatment with EHBP averted renal damage and the histologic study of kidneys showed images of tissue recovery (**Figure 3**). These results suggest that the use of EHBP may be of great value in reducing renal damage.

Bozal *et al.* [62] showed that all growth cartilage and metaphyseal bone histomorphometric parameters were significantly lower in animals exposed to UN at 48 h than in controls. EHBP administration was found to prevent this condition at 48 h reaching similar values to those of controls. Although histomorphometric values did not reach control values at 14 days, they were higher than those of animals exposed to UN at 48 h not treated with EHBP. It is noteworthy that these values also decreased in animals

Figure 3.
(HE 100X) Histological sections of cortical zone of kidneys of a control animal (A), of an animal 48 h post-intoxication with uranyl nitrate (UN) (B), and of an animal intoxicated with UN and treated with oral EHBP, 14 days post-intoxication (C). (A) Note the integrity of tubule and glomerular structure. (B) Note the marked vacuolization of the tubules, abundant hyaline cylinders, and extensive areas of necrosis, all of them usual uranium-induced lesions. (C) Note that instead the presence of scattered hyaline cylinders, the tissue shows marked recovery of renal architecture.

only receiving EHBP at 14 days. Our results show that EHBP effectively ameliorates the adverse effects of a lethal dose of UN on endochondral ossification (**Figure 4**).

In our laboratory, we have also evaluated the effect of treatment with EHBP on the reduction in interradicular bone volume and the alteration of histomorphometric parameters of bone remodeling in animals intoxicated with a lethal oral dose of UN (unpublished data). These studies showed that 48 h after UN intoxication, EHBP treatment enables an interradicular bone volume to be maintained which is similar to the controls, and this condition is sustained 14 days post-treatment (**Figure 5**). Moreover, at 48 h, EHBP prevented the reduction in bone formation and increase in bone resorption caused by UN intoxication in the interradicular bone of intoxicated animals.

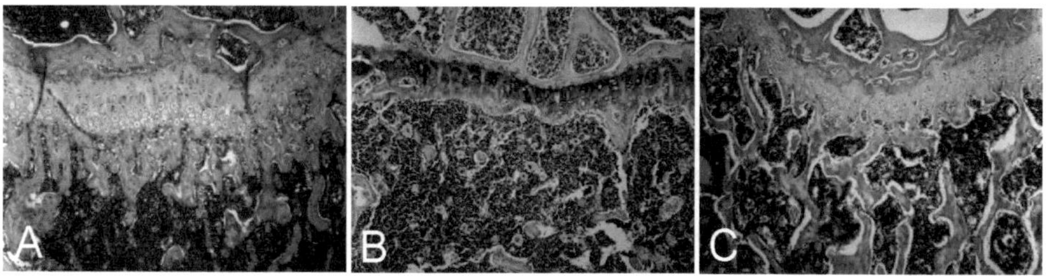

Figure 4.
(HE 100X) Histological sections of the metaphyseal bone of a control animal (A), of an animal 48 h post-intoxication with uranyl nitrate (UN) (B), and of an animal intoxicated with UN and treated with oral EHBP, 14 days post-intoxication (C). Note the marked reduction in cartilage width and the absence of proliferation cells in the growth cartilage as well as the absence of subchondral bone in UN-exposed animal (B) compared to the control (A). In animal intoxicated with UN and treated with EHBP (C), note that instead of the reduction in cartilage width compared to the control, proliferation cells in the growth cartilage and subchondral bone are evident.

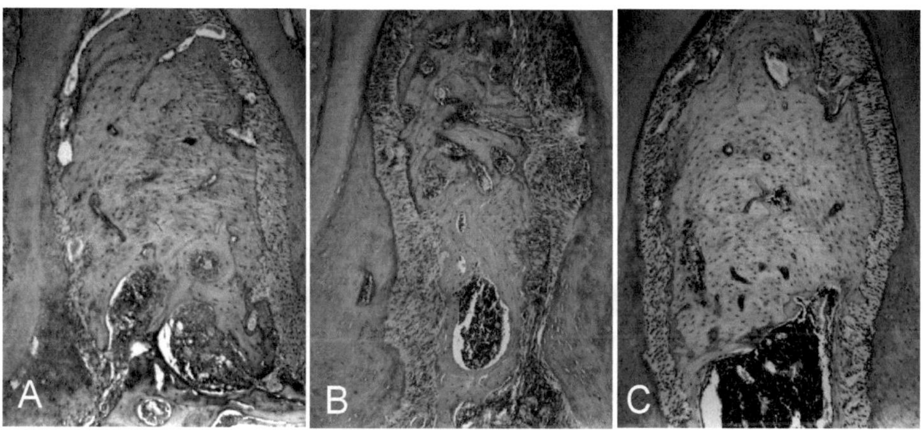

Figure 5.
(HE 100X) Histological sections of the interradicular alveolar bone of a control animal (A), of an animal 48 h post-intoxication with UN (B), and of an animal intoxicated with UN and treated with oral EHBP, 14 days post-intoxication (C). The images show the reduction in interradicular bone volume at 48 h in animals exposed to UN and how, in the exposed animals treated with EHBP, bone volume is similar to the control.

5. Discussion

Our research shows that in adult mice that had been exposed to a lethal dose of orally administered UN, a single dose of EHBP—either by mouth or subcutaneous—reduced mortality by about 50%. Surviving exposed animals had adequate renal function and showed a reduction in the deleterious effects of uranium on endochondral ossification and alveolar bone [41, 62]. We decided to administer UN by mouth so that the experimental models would replicate, as closely as possible, the situation of workers exposed to potential accidents. Given the similarity in survival rates observed with both the EHBP administration routes tested at our laboratory, we suggest that its effectiveness as a chelator to reduce the lethal effects of uranium is independent of whether it is administered orally or subcutaneously. In contrast to other studies, it is important to highlight that in the experimental design tested at our laboratory, all animals had free access to food and drink throughout, in order to recreate a situation that would similar to what might happen in case of accidental U intoxication to humans. This information reinforces the potential use of EHBP as an antidote to U, highlighting its easy accessibility for use in accidental intoxications when it is impossible to know the content of the gastrointestinal tract of the individual and which may ultimately interfere with the pharmacokinetics or pharmacological efficacy of EHBP. It is worth noting that at that time, there was no report in the literature of exposure to uranium and administration of an antidote via the same route—in our case, by mouth. EHBP was selected as uranium chelating agent based on the findings of Ubios *et al.* [72], who postulated that since bisphosphonates have proven affinity for calcium [73], they may act as U chelating agents. That study demonstrated the efficacy of only one injection of EHBP to prevent renal damage and counteract mortality due to uranium poisoning with a success rate of 100% [72].

Several studies have focused on the efficacy of different chelating agents for removing uranium from tissue deposits. However, chelation of a heavy metal is more beneficial than its removal from tissue deposits because it prevents it from reaching target organs. This property has been demonstrated experimentally in biochemical, histological, and histomorphometric studies on the kidney [41] and bones [62] of animals exposed to lethal doses of uranyl nitrate.

One favorable factor was the brief time—20 min in our studies—elapsed between the administration of U and the administration of EHBP. The time that elapses between administration of U and the antidote is critical. Catsch *et al.* [74] demonstrated that there is no apparent benefit from administrating an antidote if the time elapsed is longer than 6 h. Ubios *et al.* [75] tested the application of single doses of two different bisphosphonates acting as chelating agents—EHBP and APD (pamidronate)—observing that the animals treated with EHBP or APD up to 24 h after the exposure achieved 100% survival until the 60th day. Only when it was given 48 h after the exposure to uranium, EHBP appeared unable to prevent death. The intervals proposed by other authors range from 10 min to 24 h [67, 76, 77], with better results having been achieved by those who administered the chelating agent 10 min to 3 h post-intoxication (Tiron, in this case). It is worth noting that none of the experimental cases reported achieved a higher survival rate than we did. Some authors highlighted the importance of administrating repeated doses of the chelating agent in order to achieve a higher survival rate in animals intoxicated with U compounds [67, 76, 77] without having achieved better results than with a single dose.

6. Conclusion

The effects of a lethal dose of uranyl nitrate can be counteracted by the chelating action of bisodic etidronate administered by mouth or subcutaneously. The therapeutic effect of EHBP has been demonstrated using an animal model of uranium intoxication. Administration of EHBP provided a survival rate of 45–50%; returned serum renal biomarkers to values close to normal, which is consistent with reduction in hyalinization and necrotic areas; and reduced bone growth inhibition, reverting the damage typical of acute uranium intoxication. These results suggest that EHBP is a chelating agent capable of effectively neutralizing lethal uranium intoxication.

Acknowledgements

In Memoriam of Dr. Romulo Luis Cabrini, who was our Laboratory Director for more than 40 years and our scientific guide in the field of uranium toxicology research.

Author details

Adriana Beatriz Martínez[1*], Carola Bettina Bozal[2], Nadia Soledad Orona[3], Deborah Ruth Tasat[2,3] and Angela Matilde Ubios[2]

1 Pharmacology Department, Faculty of Dentistry, National University of Rosario, Argentina

2 Universidad de Buenos Aires, Facultad de Odontología, Cátedra de Histología y Embriología, Buenos Aires, Argentina

3 School of Science and Technology, National University of San Martin, Buenos Aires, Argentina

*Address all correspondence to: adrianabmartinez@yahoo.com.ar

References

[1] Toxicological Phealth and rofile for Uranium. U.S. Department of Health and Human Services. Agency for Toxic Substances and Disease Registry. 2013. Available from: https://www.atsdr.cdc.gov/sites/toxzine/docs/uranium_toxzine.pdf [Accessed: 07 March 2020]

[2] Clarksson T. Factors involved in heavy metals poisoning. Federation Proceedings. 2002;**36**(5):1634-1639

[3] International Comission on Radiation Protection (ICRP). Age-Dependent Doses to Members of the Public from Intake of Radionuclides: Part 4, Inhalation Dose Coefficients. ICRP Publication 71. Oxford: Pergamon Press; 1996

[4] Hursh J, Spoor N. Data on man. In: Hodge H, Stannard J, Hursh J, editors. Uranium, Plutonium Transplutonic Elements. Handbook of Experimental Pharmacology. Vol. 36. Berlin/Heidelberg: Springer Verlag; 1973. pp. 197-239

[5] Walinder G, Hammarstrom L, Billandelle U. Incorporation of uranium. I. Distribution of intravenously and intraperitoneally injected uranium. British Journal of Industrial Medicine. 1967;**24**:305-312. DOI: 10.1136/oem.24.4.305

[6] La Touche Y, Willis D, Dawydiak O. Absorption and biokinetics of U in rats following an oral administration of uranyl nitrate solution. Health Physics. 1987;**53**:147-162. DOI: 10.1097/00004032-198708000-00005

[7] Durbin P, Wrenn M. Metabolism and effects of uranium in animals. In: Wrenn M, editor. Conference: Occupational Health Experience with Uranium, Arlington, VA. Vol. 93. Washington, DC: U.S. Energy Research and Development Administration (ERDA); 1975. pp. 67-129

[8] Sullivan M, Ruemmler P, Ryan J, Buschbom R. Influence of oxidizing or reducing agents on gastrointestinal absorption of U, Pu, Am, Cm and Pm by rats. Health Physics. 1986;**50**(2):223-232. DOI: 10.1097/00004032-198602000-00006

[9] Vidaud C, Dedieu A, Basset C, Plantevin S, Dany, Olivier Pible I, et al. Screening of human serum proteins for uranium binding. Chemical Research in Toxicology. 2005;**18**(6):946-953. DOI: 10.1021/tx050038v

[10] Wedeen R. Renal diseases of occupational origin. Occupational Medicine. 1992;**7**(3):449-463

[11] Hursh J, Neuman W, Toribara T, Wilson H, Waterhouse C. Oral ingestion of uranium by man. Health Physics. 1969;**17**:619-621

[12] Orcutt J. The toxicology of compounds of uranium following application to the skin. In: Voegtlin C, Hodge HC, editors. Pharmacology and Toxicology of Uranium Compounds, Vols. 3 and 4. New York, NY: McGraw Hill Book Co.; 1949. pp. 377-414

[13] de Rey B, Lanfranchi H, Cabrini R. Percutaneous absorption of uranium compounds. Environmental Research. 1983;**30**:480-491. DOI: 10.1016/0013-9351(83)90233-5

[14] Harris BW et al. Experimental clearance of urannium dust from the body. In: Darries CN, editor. Inhaled Particles and Vapours. London, UK: Pergamon; 1961

[15] Kurttio P, Harmoinen A, Saha H, Salonen L, Karpas Z, Komulainen H, et al. Kidney toxicity of ingested uranium from drinking water. American Journal of Kidney Diseases. 2006;**47**(6):972-982. DOI: 10.1053/j.ajkd.2006.03.002

[16] Kathren R, Burklin R. Acute chemical toxicity of uranium. Health Physics. 2008;**94**(2):170-179. DOI: 10.1097/01.HP.0000288043.94908.1f

[17] Harrison J, Stather J. The gastrointestinal absorption of protactinium, uranium, and neptunium in the hamster. Radiation Research. 1981;**88**(1):47-55

[18] Anke M, Seebera O, Muller R, Schafer U, Zerull. Uranium transfer in the food chain from soil to plants, animals and man. Chemie der Erde-Geochemistry. 2009;**69**:75-90. DOI: 10.1016/j.chemer.2007.12.001

[19] Zamora M, Tracy B, Zielinski J, Meyerhof D, Moss M. Chronic ingestion of uranium in drinking water: A study of kidney bioeffects in humans. Toxicological Sciences. 1998;**43**(1): 68-77. DOI: 10.1006/toxs.1998.2426

[20] WISE Uranium Project. 1999. Available from: www.wiseproject.org [Accessed: 20 December 2019]

[21] Swanson S. Food chain transfer of U-series radionuclides in northern Saskatchewan aquatic system. Health Physics. 1985;**49**:747-770. DOI: 10.1097/00004032-198511000-00008

[22] Lapham S, Millard J, Samet J. Health implications of radionuclide levels in cattle raised near uranium mining and milling facilities in Ambrosia Lake, New Mexico. Health Physics. 1989;**36**(3):327-330. DOI: 10.1097/00004032-198903000-00008

[23] EPA-US Environmental Protective Agency Code of Federal Regulation CFR 421, Subpart AD; 1985

[24] US Department of Health & Human Services. Public Health Service. Agency for Toxic Substances and Disease Registry. Research Triangle Institute "Uranium: Draft for Public Comment"; 1997. Bookshelf ID: NBK158802PMID: 24049861

[25] Maynard E, Down W, Hodge H. Oral toxicity of uranium compounds. In: Voegtlin C, Hodge H, editors. Pharmacology and Toxicology of Uranium Compounds. New York, NY: McGraw-Hill; 1953

[26] Martinez A, Cabrini R, Ubios A. Orally administered ethane-1-hydroxy-1,1-bisphosphonate reduces the lethal effect of oral uranium poisoning. Health Physics. 2000;**78**:668-671. DOI: 10.1097/00004032-200006000-00009

[27] Dublineau I, Grison S, Baudelin C, Dudoignon N, Souidi M, Marquette C, et al. Absorption of uranium through the entire gastrointestinal tract of the rat. International Journal of Radiation Biology. 2005;**81**(6):473-482. DOI: 10.1080/09553000500196029

[28] Dublineau I, Grison S, Linard C, Baudelin C, Dudoignon N, Souidi M, et al. Short-term effects of depleted uranium on immune status in rat intestine. Journal of Toxicology and Environmental Health. Part A. 2006;**69**(17):1613-1628. DOI: 10.1080/15287390600629825

[29] Dublineau I, Grandcolas L, Grison S, Baudelin C, Paquet F, Voisin P, et al. Modifications of inflammatory pathways in rat intestine following chronic ingestion of depleted uranium. Toxicological Sciences. 2007;**98**(2): 458-468. DOI: 10.1093/toxsci/kfm132

[30] Tasat DR, de Rey BM. Cytotoxic effect of uranium dioxide on rat alveolar macrophages. Environmental Research. 1987;**44**:71-81. DOI: 10.1016/s0013-9351(87)80087-7

[31] Orona NS, Tasat DR. Uranyl nitrate-exposed rat alveolar macrophages cell death: Influence of superoxide anion and TNF α mediators. Toxicology and Applied Pharmacology. 2012;**261**(3):309-316. DOI: 10.1016/j.taap.2012.04.022

[32] Katz SA. The chemistry and toxicology of depleted uranium. Toxics. 2014;**2**:50-78. DOI: 10.3390/toxics2010050 toxics. ISSN: 2305-6304

[33] Craft E, Abu-Qare A, Flaherty M, Garofolo M, Rincavage H, Abou-Donia M. Depleted and natural uranium: chemistry and toxicological effects. Journal of Toxicology and Environmental Health. Part B, Critical Reviews. 2004;**7**(4):297-317. DOI: 10.1080/10937400490452714

[34] Briner W. The toxicity of depleted uranium. International Journal of Environmental Research and Public Health. 2010;**7**:303-313. DOI: 10.3390/ijerph7010303

[35] López R, Díaz Sylvester P, Ubios A, Cabrini R. Percutaneous toxicity of uranyl nitrate: Its effect in terms of exposure area and time. Health Physics. 2000;**78**(4):434-437. DOI: 10.1097/00004032-200004000-00007

[36] McDiarmid MA, Keogh JP, Hooper FJ, McPhaul K, Squibb K, Kane R, et al. Health effects of depleted uranium on exposed Gulf War veterans. Environmental Research. 2000;**82**(2):168-180. DOI: 10.1006/enrs.1999.4012

[37] de Rey B, Lanfranchi H, Cabrini R. Deposition pattern and toxicity of subcutaneously implanted uranium dioxide in rats. Health Physics. 1984;**46**:688-692

[38] Voegtlin C, Hodge H. Pharmacology and Toxicology of Uranium Compounds (First edition). New York/London: McGraw Hill; 1949

[39] Stokinger H, Baxter R, Dygert H, et al. Toxicity following inhalation for 1 and 2 years. In: Voegtlin C, Hodge HC, editors. Pharmacology and Toxicology of Uranium Compounds. New York, NY: McGraw-Hill; 1953. pp. 1370-1776

[40] Domingo J, Llobet J, Tomás J, Corbella J. Acute toxicity of uranium in rats and mice. Bulletin of Environmental Contamination and Toxicology. 1987;**39**:168-174. DOI: 10.1007/bf01691806

[41] Martínez A, Mandalunis P, Bozal C, Cabrini R, Ubios A. Renal function in mice poisoned with oral uranium and treated with ethane-1-hydroxy-1,1-bisphosphonate (EHBP). Health Physics. 2003;**85**:343-347. DOI: 10.1097/00004032-200309000-00010

[42] Hodge H. Mechanism of uranium poisoning. In: Proceedings of the International Conference of the Peaceful Uses of Atomic Energy, Vol. 13; 1955. pp. 229-232

[43] Avasthi P, Evan A, Hay D. Glomerular endothelial cells in uranyl nitrate-induced acute renal failure in rats. The Journal of Clinical Investigation. 1980;**65**(1):121-127. DOI: 10.1172/JCI109641

[44] Haley D. Morphologic changes in uranyl nitrate-induced acute renal failure in saline- and water-drinking rats. Laboratory Investigation. 1982;**46**(2):196-208

[45] Haley D, Bulger R, Dobyan D. The long-term effects of uranyl nitrate on the structure and function of the rat kidney. Virchows Archiv. B, Cell Pathology Including Molecular Pathology. 1982;**41**(1-2):181-192. DOI: 10.1007/bf02890280

[46] Kobayashi S, Nagase M, Honda N, Hishida A, et al. Kidney International. 1984;**26**(6):808-815. DOI: 10.1038/ki.1984.222

[47] Sangeetha Vijayan P, Rekha P, Dinesh U, Arun A. Biochemical and histopathological responses of the Swiss albino mice treated with uranyl nitrate and its recovery. Renal

Failure. 2016;**38**(5):770-775. DOI: 10.3109/0886022X.2016.1160248

[48] Adams N, Spoor N. Kidney and bone retention functions in the human metabolism of uranium. Physics in Medicine and Biology. 1974;**19**:460-471. DOI: 10.1088/0031-9155/19/4/004

[49] Neuman M, Neuman W. The deposition of uranium in bone; radioautographic studies. The Journal of Biological Chemistry. 1948;**175**(2):711-714

[50] Rowland R, Farnham J. The deposition of uranium in bone. Health Physics. 1969;**17**:139-144. DOI: 10.1097/00004032-196907000-00015

[51] Kurttio P, Komulainen H, Leino A, Salonen L, Auvinen A, Saha H. Bone as a possible target of chemical toxicity of natural uranium in drinking water. Environmental Health Perspectives. 2005;**113**:68-72. DOI: 10.1289/ehp.7475

[52] Neuman W, Neuman M, Mulryan B. The deposition of uranium in bone; animal studies. The Journal of Biological Chemistry. 1948;**175**(2):705-709

[53] Guglielmotti M, Ubios A, de Rey B, Cabrini R. Effects of acute intoxication with uranyl nitrate on bone formation. Experientia. 1984;**40**:474-476. DOI: 10.1007/bf01952392

[54] Guglielmotti M, Ubios A, Cabrini R. Alveolar wound healing alterations under uranyl nitrate intoxication. Journal of Oral Pathology. 1985;**14**:565-572. DOI: 10.1111/j.1600-0714.1985.tb00530.x

[55] Guglielmotti M, Ubios A, Cabrini R. Morphometric study of the effect of low dose of uranium on bone healing. Acta Stereologica. 1987;**6**:357-366

[56] Ubios A, Marzorati M, Cabrini R. Ultraestructural alterations of bone due to uranium intoxication. Journal of Dental Research. 1992;**71**:973

[57] Ubios A, Braun E, Cabrini R. Lethality due to uranium poisoning is prevented by ethane-1-hydroxy-1,1-bisphosphonate (EHBP). Health Physics. 1994;**66**:540-544. DOI: 10.1097/00004032-199405000-00005

[58] Ubios AM, Piloni MJ, Marzorati M, Cabrini RL. Bone growth is impaired by uranium intoxication. Acta Odontológica Latinoamericana. 1995;**8**:3-8

[59] Pujadas Bigi M, Lemlich L, Mandalunis P, Ubios A. Exposure to oral uranyl nitrate delays tooth eruption and development. Health Physics. 2003;**84**(2):163-169. DOI: 10.1097/00004032-200302000-00003

[60] Tasat D, Orona N, Mandalunis P, Cabrini R, Ubios A. Ultrastructural and metabolic changes in osteoblasts exposed to uranyl nitrate. Archives of Toxicology. 2007;**81**(5):319-326. DOI: 10.1007/s00204-006-0165-2

[61] Ubios A, Guglielmotti M, Steimetz T, Cabrini R. Uranium inhibits bone formation in physiologic alveolar bone modeling and remodeling. Environmental Research. 1991;**54**:17-23. DOI: 10.1016/s0013-9351(05)80191-4

[62] Bozal C, Martinez A, Cabrini R, Ubios A. Effect of ethane-1-hydroxy-1,1- bisphosphonate (EHBP) on endochondral ossification lesions induced by a lethal oral dose of uranyl nitrate. Archives of Toxicology. 2005;**79**:475-481. DOI: 10.1007/s00204-005-0649-5

[63] Gritsaenko T, Pierrefite-Carle V, Lorivel T, Breuil V, Carle GF, Santucci-Darmanin S. Natural uranium impairs the differentiation and the

resorbing function of osteoclasts. Biochimica et Biophysica Acta - General Subjects. 2017;**1861**(4):715-726. DOI: 10.1016/j.bbagen.2017.01.008

[64] Lawrence G, Patel K, Nusbaum A. Uranium toxicity and chelation therapy. Conference paper. Pure and Applied Chemistry. 2014;**86**(7):1105-1110. DOI: 10.1515/pac-2014-0109

[65] Cooper JR, Stradling GN, Smith H, Ham SE. The behaviour of uranium-233 oxide and uranyl-233 nitrate in rats. International Journal of Radiation Biology and Related Studies in Physics, Chemistry, and Medicine. 1982;**41**(4):421-433. DOI: 10.1080/09553008214550461

[66] Fisher D, Kathern R, Swint M. Modified biokinetic model for uranium from analysis of acute exposure of UF6. Health Physics. 1991;**60**(3):335-342. DOI: 10.1097/00004032-199103000-00002

[67] Domingo J, Colomina M, Llobet J, Jones M, Singh P, Campbell R. The action of chelating agents in experimental uranium intoxication in mice: Variation with structure and time of administration. Fundamental and Applied Toxicology. 1992;**19**:350-357. DOI: 10.1016/0272-0590(92)90173-f

[68] Ortega A, Domingo J, Gomez M, Corbella J. Treatment of experimental acute uranium poisoning by chelating agents. Pharmacology and Toxicology. 1989;**64**:247-251. DOI: 10.1111/j.1600-0773.1989.tb00640.x

[69] Stradling G, Gray S, Moody J, Ellender M. Efficacy of Tiron for enhancing the excretion of uranium from the rat. Human & Experimental Toxicology. 1991;**10**(3):195-198. DOI: 10.1177/096032719101000308

[70] Ubios A, Guglielmotti M, Cabrini R. Effects of diphosphonates on the prevention of X radiation induced inhibition of bone formation in rats. Journal of Oral Pathology. 1986;**15**:500-505. DOI: 10.1111/j.1600-0714.1986.tb00666.x

[71] Guglielmotti M, Ubios A, Larumbe J, Cabrini R. Tetracycline in uranyl nitrate intoxication: Its action on renal damage and uranium retention in bone. Health Physics. 1989;**57**:403-405. DOI: 10.1097/00004032-198909000-00005

[72] Ubios A, Guglielmotti M, Cabrini R. Ethane 1-hydroxy-1, 1-diphosphonate (EHDP) counteracts the inhibitory effect of uranyl nitrate on bone formation. Archives of Environmental Health. 1990;**45**(6):374-377. DOI: 10.1080/00039896.1990.10118758

[73] Fleisch H, Bisaz S. Isolation from urine of pyrophosphate, a calcification inhibitor. The American Journal of Physiology. 1962;**203**:671-675. DOI: 10.1152/ajplegacy.1962.203.4.671

[74] Catsch A. Effect of some chelating agents on acute toxicity of uranyl nitrate. Klinische Wochenschrift. 1959;**37**(12):657-660. DOI: 10.1007/bf01478404

[75] Ubios A, Braun EM, Cabrini R. Effect of bisphosphonates on abnormal mandibular growth of rats intoxicated with uranium. Health Physics. 1998;**75**(6):610-613. DOI: 10.1097/00004032-199812000-00004

[76] Basinger MA, Jones MM. Tiron (sodium 4,5-dihydroxybenzene-1,3-disulfonate) as an antidote for acute uranium intoxication in mice. Research Communications in Chemical Pathology and Pharmacology. 1981;**34**(2):351-357

[77] Basinger MA, Forti RL, Burka LT, Jones MM, Mitchell WM, Johnson JE, et al. Phenolic chelating agents as antidotes for acute uranyl acetate intoxication in mice. Journal of Toxicology and Environmental Health. 1983;**11**:237-246

Chapter 9

Influence of the Doping Ion Nature and Content on Defect Creation Processes under the Effect of Ionizing Radiation in Aluminoborosilicate Glasses

Eugenia Malchukova

Abstract

Effects of ionizing irradiation on defect creation processes have been studied in rare earth (RE)-doped (RE = Sm, Gd, Eu, Ce, Nd) aluminoborosilicate glass with use of the electron paramagnetic resonance (EPR) and optical spectroscopy. As a function of RE ion nature, we observe that doping significantly influences the nature of the defects produced during irradiation and more specifically the relative proportions between hole and electron defect centers. Strong decrease of defect production efficiency under ionizing radiation independence on both the RE doping content and on the relative stability of the RE different oxidation states is also clearly revealed. The results could be explained by dynamical reversible trapping of the electron-hole pairs produced during irradiation on the different RE charge states as well as by RE segregation and pre-existing defects speciation in ABS glass structure.

Keywords: borosilicate, glasses, EPR, luminescence, irradiation, defects

1. Introduction

Irradiation effects are an active research field in amorphous silica (aSiO2) due to many technological applications requiring a good maintenance of transparency (e.g. fibers, laser optics and radioactive environments) [1–4]. Indeed for aSiO2, the optical properties are controlled by the nature and the content of defect produced during an ionizing radiation (laser, X, γ, electrons) [5–7]. Different works using Electron Paramagnetic Spectroscopy (EPR) and optical absorption have shown that two different defects production processes occur during ionizing irradiation. The first process called intrinsic defects production [8–10] is correlated to Si–O bonds breaking leading to well-known paramagnetic E′ and Non-Bridging Oxygen Hole Centers (NBOHC) with self-trap excitons acting as possible precursors [11–13]. Peroxy radicals (POR) paramagnetic defect can also be produced with intrinsic process by the displacement of oxygen into an interstitial position like Frenkel defects [12, 14]. The second defect creation process is called "extrinsic" and is correlated to the presence of different impurities (H, Cl, Transition metals, Rare earth, ...) inside aSiO2 materials [8, 15–18]. In that case, the nature of different possible irradiation paramagnetic defects produced (E′, NBOHC, and POR) does not differ but an higher defect production efficiency is observed associated to saturation processes of defect content depending on impurities nature and content, respectively [19]. Defects production processes in aSiO2 are therefore mainly controlled by the nature and the content of the different impurities.

For more complex oxide glass compositions, the presence of network modifiers ions (Na^+, K^+, Ca^{2+}) and other network formers ions (B^{3+}, Al^{3+}) stabilizes with an high efficiency different hole trap like Boron-Oxygen Hole Center (BOHC) [20, 21], Aluminum-Oxygen Hole Center (AlOHC) [22] or Hole trapped defects on Non-Bridging Oxygen (NBO) called HC1, HC2 centers [23, 24]. In addition, the literature shows that the content of electron trap defects like E′ centers and equivalent defects closed to B^{3+} (BEC center) and Al^{3+} ions are generally much more lower than hole trap defects content for all oxide glass compositions including silica. Electron trapping on glass impurities like Hydrogen, alkaline, Transition Metals (TM), or Rare Earth (RE) ions could explain differences between the content of hole and electron trapped defects produced during exposure to ionizing radiation [7, 15]. In general for oxide glasses like borosilicate, silicate, and aluminosilicate, the nature and content of different paramagnetic defects observed by EPR spectroscopy will depend on the relative proportion of different network formers (Si^{4+}, Al^{3+}, and B^{3+}) and on network modifiers contents introduced inside the oxide glass. In case of aluminoborosilicate (ABS) glasses studied in this work, the nature and proportion of different paramagnetic defects have been previously determined using the simulation of EPR spectra of different β-irradiated borosilicate glass samples [25].

Doping processes of oxide glasses with rare earth (RE) ions influence a lot the nature and the content of different paramagnetic defects produced during exposure to ionizing radiation. In case of Sm- [26], Gd- [27], and Yb- [28] doped borosilicate glass compositions, the first effect of doping is the decrease of total paramagnetic defect contents produced during irradiation to one integrated dose. Moreover for Fe-doped soda-lime glasses [29] and Cr-doped silicate glasses [30], a complete disappearance of the different paramagnetic defects is observed for doping level around 1 mol. %. Associated to the decrease of paramagnetic defects production efficiency with the doping ion content, the decrease of different structural changes under irradiation detected by Raman spectroscopy [29, 30], (increase of polymerization and the molecular oxygen production, and decrease of Si–O–Si average angle) are also observed at integrated dose higher than 10^9 Gy. Structural changes in glasses under the effect of ionizing radiation are mainly controlled by alkaline mobility in both network modifiers and charge compensator positions [31]. This result shows therefore strong relationships between the nature and contents of doping ions, irradiation defects creation processes, and the structural changes due to ionizing radiation exposure in oxide glasses.

However, these previous studies have mainly focused on the modification of the total paramagnetic defect concentration produced during ionizing radiation as a function of the doping ion nature and content. The goal of this chapter is to systematically present the paramagnetic irradiation defect creation processes in rare earth-doped oxide glasses. First, the influence of RE doping ion nature on the relative paramagnetic defect proportion observed by EPR spectroscopy in the same ABS glass composition will be considered. Then, the influence of both doping ion content and integrated radiation dose on the nature, content, and relative proportion of the different paramagnetic defects produced during ionizing radiation will be considered also. It is known, that the optical spectroscopy is most useful in cases where EPR techniques are not applicable and for diamagnetic defects. Also, transmission and luminescence experiments will be carried out in order to provide additional information on RE doping effect on ABS glass structure. This approach could improve our knowledge about defect creation processes under irradiation for ion doped oxide glasses. For that purpose, different RE-doped ABS glasses (RE = Sm, Gd, Eu, Ce, and Nd where the RE doping level between 0.1 and 1 mol. %) have been irradiated to different doses between 10^5 and 2×10^9 Gy using a Van de Graaff accelerator.

2. Experimental part

Rare earth-doped ABS glasses were prepared by adding to the base glass with the following composition—59.77% SiO_2, 4.00% Al_2O_3, 22.41% B_2O_3, 12.12% Na_2O, and 1.70% ZrO_2 (in mol. %)—different amounts of doping ions. The doping content of RE oxide considered in this work is shown in **Table 1**. The dried mixed powders were heated at 750°C for 10 h in air in a Pt crucible and melted at 1500°C for 2 h, then quenched on a copper plate. Before cutting, annealing at 500°C for 1–2 h was necessary to remove the internal stress. Samples were polished on a hand grinding wheel with a silicon carbide abrasive having the average grain size of 10 μm (1000 grain) to achieve the average thickness of 0.56 ± 0.05 mm. Each glass was analyzed by X-ray diffraction, in order to confirm the amorphous characteristics of the glass.

All glasses were β-irradiated with 2.5 MeV electrons (10 μA) provided by a Van de Graaff accelerator (LSI, Palaiseau, France) at different integrated doses from 10^6 to 2×10^9 Gy. The used sample thickness made it possible to obtain uniform irradiation on the entire glass volume. EPR measurements were conducted at room temperature on a X band (ν = 9.420 GHz) EMX Brücker EPR spectrometer using a 100 kHz field modulation, 3×10^{-4} Tesla of amplitude modulation and an applied microwave power of 1 mW. The EPR spectra of all irradiated RE-doped ABS glasses have been normalized to the same receiver gain and to a 100 mg sample weight. Paramagnetic defects total content has been estimated by the area under the defect absorbance EPR spectrum. A maximum error of 10% has been considered in this work taking into account uncertainties in the irradiated glass samples weight measurement, the sample positioning inside spectrometer cavity and defect absorbance EPR spectrum area computation or EPR line intensity measurement. The optical transmission spectra were measured on an Agilent Varian Cary 5000 spectrophotometer in 1 nm steps in the range of 200–1500 nm. The photoluminescence was analyzed by a SHAMROCK spectrograph F5303 mm: 150 lines/mm grating and a 400 mm slit combined with an ANDOR Istar (Andor Company, Belfast, U.K.) intensified charge coupled device. The 266 nm wavelength pulses width of around 8 ns and laser repetition rate of 10 Hz of an INDI Nd:YAG pulsed laser spectra physics were used for the PL excitation. The laser beam is transported via two mirrors, two lenses, and three diaphragms to the sample center with a final diameter of 2 mm. The pulse

Rare earth	mol%
Gd_2O_3	0.1; 0.2; 0.5; 1
Sm_2O_3	0.1; 0.2; 0.5; 1
Eu_2O_3	0.1; 0.2; 0.5; 1
Nd_2O_3	0.1; 0.2; 0.6; 1
CeO_2	0.2; 0.4; 1.2; 2

Table 1.
RE doping concentration.

energy on the sample was ~2 mJ/pulse. The spectral measurements were carried out using different delay time (d) and gate width (G).

3. EPR spectra of RE-doped ABS glass: b-irradiation dose effect

Without doping, the nature of the different defects produced by ionizing radiation has been previously studied in the non-doped ABS glass composition irradiated with 2.5 MeV electrons by the simulation of the EPR spectra of irradiated samples annealed at different temperatures [25]. The EPR of the non-doped ABS glass irradiated at 1.3×10^8 Gy is presented in **Figure 1**. The main component of these EPR spectra is associated to the hyperfine structure with 11B (I = 3/2). The defect is called the Boron-Oxygen Hole Center (BOHC) (g1 = 2.0029, g2 = 2.0115, and g3 = 2.0500) and it is attributed to a hole trap on an oxygen link to a boron atom [21]. The second hole center for this glass composition that can be observed at high annealing temperature is the peroxy radical (Si–O–O°) named Oxy defect (g1 = 2.0024, g2 = 2.0110, and g3 = 2.0439) in the literature for silicate glasses [22]. The last hole center determined by the simulation of the EPR spectra for this glass composition is the HC1 center attributed to a hole trapped on a non-bridging oxygen in the vicinity of alkaline ion. Finally, the EPR line around g = 2.0011 is an electron trap and is attributed to the well-known E′ center [11].

Figure 1 presents the EPR spectra recorded at room temperature of the non-doped and 0.1 mol% RE-doped ABS glasses (RE = Sm, Gb, Eu, Nd, and 0.2 mol% Ce) irradiated with 2.5 MeV electrons (integrated dose of 1.3×10^8 Gy). First, It can be observed

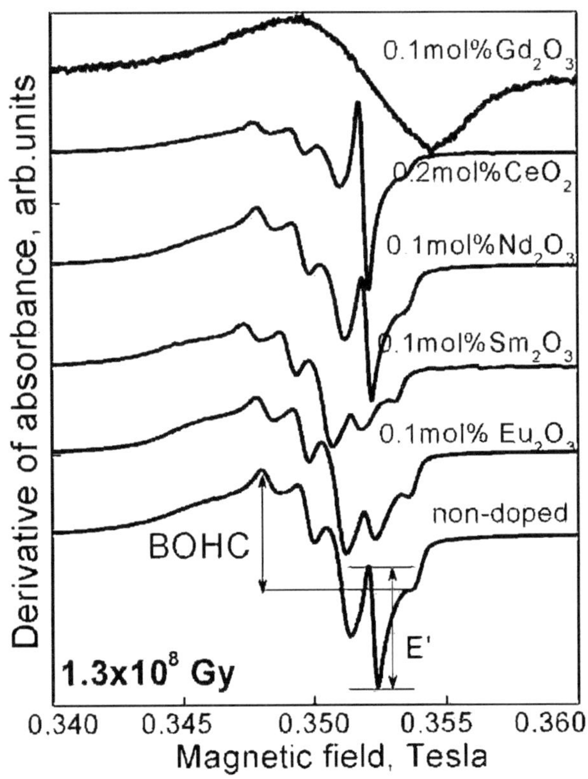

Figure 1.
X band EPR spectra of the lowest RE-doped ABS glasses (RE = Gd, Sm, Eu, Ce, and Nd) irradiated at 1.3×10^8 Gy.

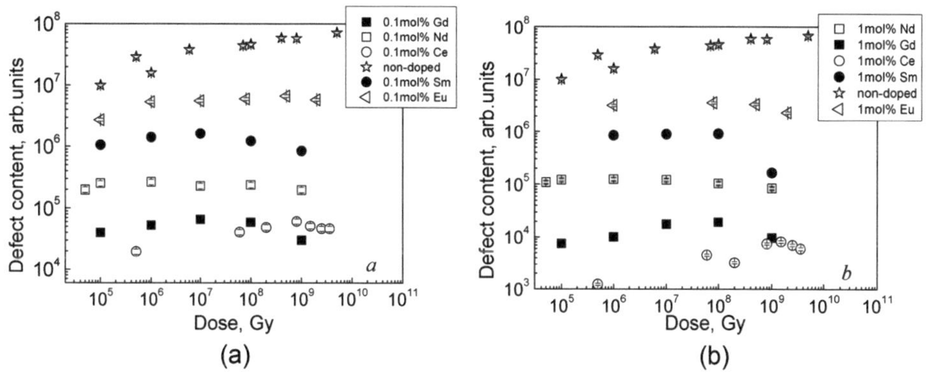

Figure 2.
Evolution of EPR defects content as a function of integrated dose for the lowest: (a) and the highest; (b) concentration of RE doping of β–irradiated ABS glass (RE = Gd, Sm, Eu, Ce, and Nd).

clearly on that figure that RE doping at low level (0.1 mol%) influences significantly both the amount of defect produced during irradiation and their relative proportions. More specifically, a decrease of the different hole centers (HC) is analyzed depending on the nature of the doping ion ([HC]Eu > [HC]Sm > [HC]Nd > [HC]Ce). This effect is maximum for 0.2 mol% Ce-doped glasses where the content of hole centers is drastically decreasing in comparison with the non-doped glass composition. For E′ center detected at g = 2.0011 in the non-doped ABS glass, a strong increase as a function of the RE doping ion nature is observed ([E′]Eu < [E′]Sm < [E′]Nd < [E′]Ce). However, for all ABS glass compositions studied, the quantity of electron defect centers (E′) remains smaller than the hole defect centers (BOHC, OXY, HC_1) showing therefore the presence of other mechanisms acting as traps for the electrons produced during irradiation.

Evolution of the total paramagnetic defect content for all RE-doped glasses (RE = Sm, Gd, Eu, Nd, and Ce) as a function of the integrated dose is presented in logarithmic scales on **Figure 2** for two different RE doping levels: 0.1 and 1 mol%, on **Figure 2a** and **b**, respectively.

For non-doped ABS glass, the total defect concentration is increasing with the integrated dose. This behavior can be correlated with extrinsic and/or intrinsic defect creation processes under the effect of ionizing radiation [8]. Two effects can be observed in **Figure 2a** and **b**. First, the decrease of the defect produced during irradiation at one integrated dose depending on the nature of the RE doping ion. The second effect is saturation behavior of the defect content as a function of the integrated dose associated to a decrease of the defect content at higher doses for all RE-doped glasses. We can therefore conclude that glass doping processes play an important role on defect creation processes under the effect of ionizing radiation.

4. EPR spectra of RE-doped ABS glass: RE concentration effect

The influence of RE content on defect production under irradiation is shown in **Figure 3a** and **b** for two different integrated doses. For all RE doping considered in this work, the defect content is decreasing as a function of the RE doping content inside the glass but with different efficiency depending on the RE nature.

In addition, the relative proportions between the different paramagnetic defects observed by EPR spectroscopy are modified by both the nature of the RE ion and also its content in the host glasses. These effects can be seen on **Figures 4–7** showing the normalized EPR spectra of irradiated (1.3×10^8 Gy) glass samples doped with

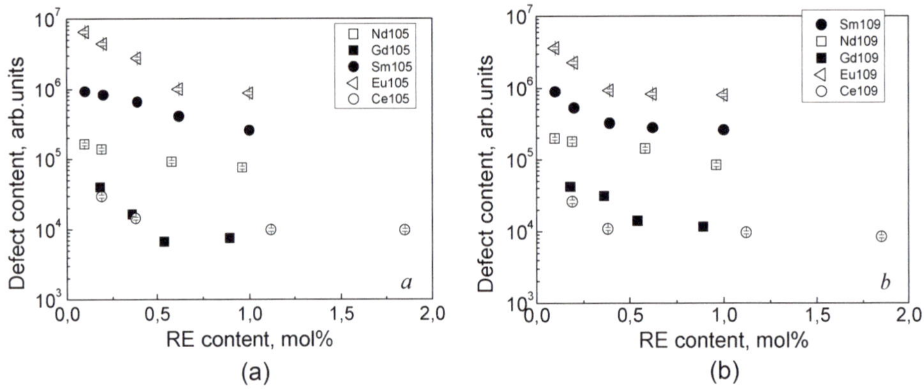

Figure 3.
Evolution of EPR defects concentration as a function of the RE doping ion content in ABS glasses β–irradiated at 6.5 × 10^{6}*: (a) and 2.6 × 10*^{9} *Gy; (b) RE = Gd, Sm, Eu, Ce, and Nd.*

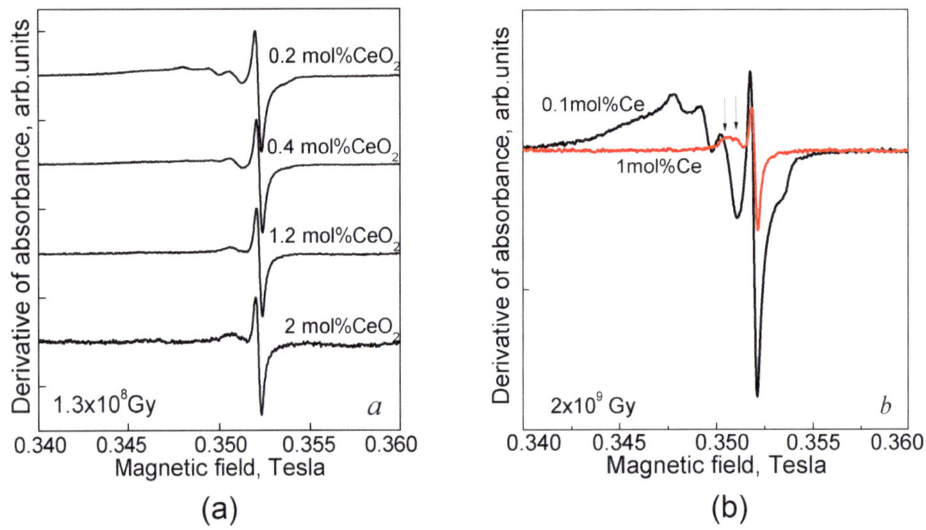

Figure 4.
X band EPR spectra recorded at room temperature of 0.2, 0.4, 1.2, and 2 mol.% CeO_2*-doped ABS glass irradiated at 1.3 × 10*^{8} *Gy (the EPR spectra have been normalized to E' EPR line intensity): (a) and X band EPR spectra of 0.2 and 2 mol.%*CeO_2*-doped ABS glass irradiated at 2 × 10*^{9} *Gy (b).*

Ce, Eu, Sm, and Nd ions, respectively. From the **Figure 4a** one can see the strongest influence of doping on defect production in Ce-doped glasses: a huge decreased of different holes centers (BOHC, OXY, HC_1) in the defect EPR spectrum is observed starting from the lowest doping level considered in this work (0.2 mol.% of CeO_2). The BOHC center becomes undetectable in the EPR spectra at Cerium doping levels higher than 0.2 mol.% and the hole defects remaining in the EPR spectra is the OXY and HC1 centers as it is shown in **Figure 4b** for the Ce-doped ABS glass irradiated at 2 × 10^9 Gy (arrows in **Figure 4b**). This effect is more pronounced in the case of the highest doses (more than 10^9 Gy). From **Figure 5a** and **b**, it can be concluded that Ce doping strongly inhibits the defect production observed by EPR spectroscopy and in addition stops the different holes defects production under the effect of ionizing radiation. BOHC defect is also detected.

When comparing these results with Eu-doping in the same ABS glass composition, similar effects of doping ion content on defect production efficiency are observed by EPR spectroscopy (**Figure 5**). But in the case of Eu-doping, a

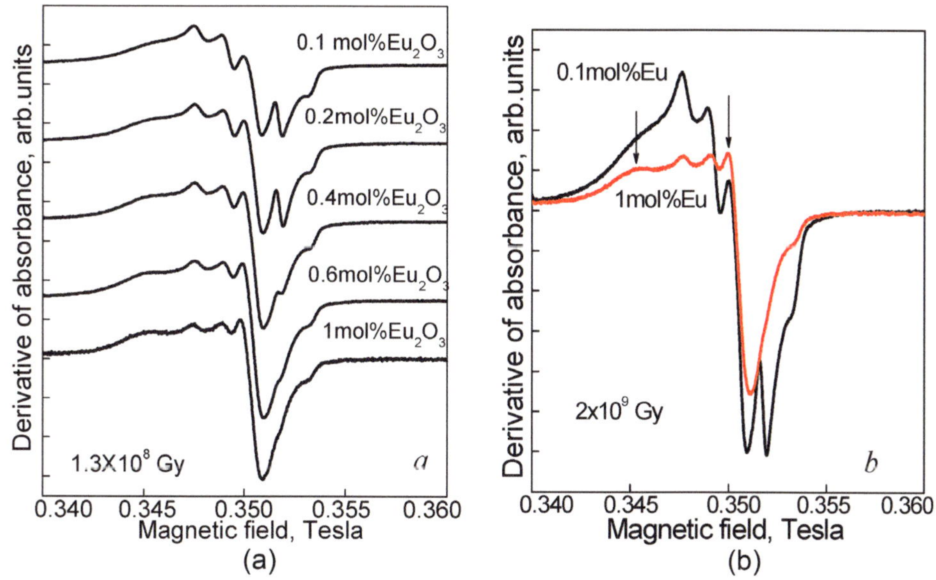

Figure 5.
X band EPR spectra recorded at room temperature of 0.1, 0.2, 0.5, and 1 mol.% Eu_2O_3-doped ABS glass irradiated at 1.3×10^8 Gy (the EPR spectra have been normalized to BOHC EPR line intensity) (a) and X band EPR spectra of 0.1 and 1 mol.% Eu_2O_3-doped ABS glass irradiated at 2×10^9 Gy (b).

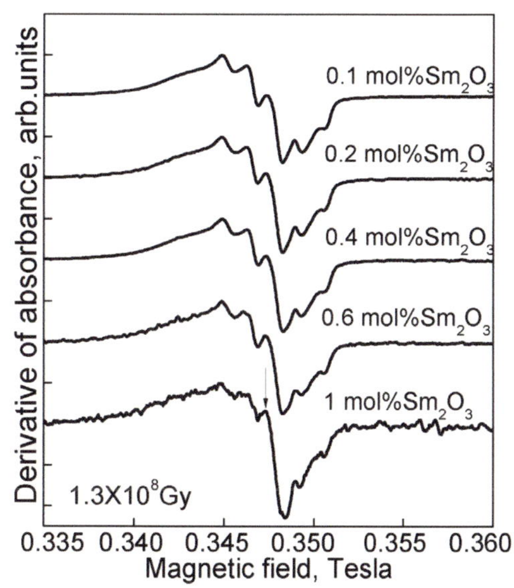

Figure 6.
X band EPR spectra recorded at room temperature of 0.1, 0.2, 0.4, 0.6, and 1 mol.% Sm_2O_3-doped ABS glass irradiated at 1.3×10^8 Gy (the EPR spectra have been normalized to BOHC EPR line intensity).

strong decrease of E′ centers contribution in the defect EPR spectra can be seen (**Figure 5a**). The E′ center becomes undetectable in the EPR spectra at Europium doping levels higher than 0.6 mol.% and the hole defects re-arrangement in the EPR spectra between OXY and BOHC centers can be observed. The highest irradiation dose results in more re-arrangement of hole defects as can be seen from **Figure 5b** (arrows in the **Figure 5b**). The influence of Sm- (**Figure 6**) and Nd- (**Figure 7**) doping contents are weaker than for Ce-doped ABS glasses for both the decrease in

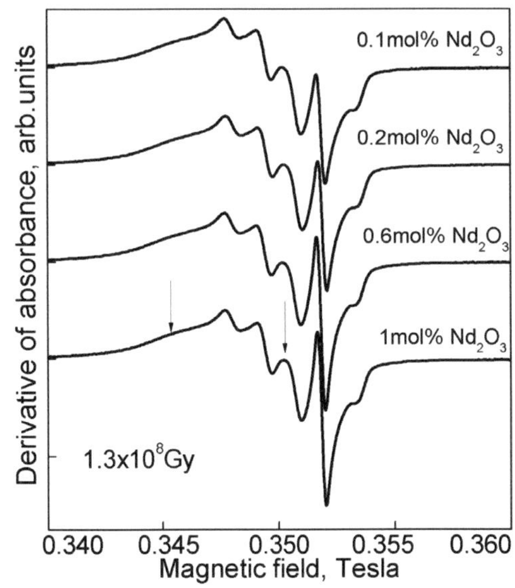

Figure 7.
X band EPR spectra recorded at room temperature of 0.1, 0.2, 0.6, and 1 mol.% Nd_2O_3-doped ABS glass irradiated at 1.3 × 10^8 Gy (the EPR spectra have been normalized to BOHC EPR line intensity).

defect production efficiencies and the changes in the relative proportions of defects composing the EPR spectra of these irradiated glass samples. However, a decrease in the relative proportion of the different defect holes centers (BOHC, Oxy, HC_1) for the different Nd-doped ABS glasses (**Figure 7**) can be observed.

This effect could also be correlated to the evolution analyzed for Ce-doped ABS glasses. The changes of the relative proportions of defect as a function of Sm-doping content in ABS glasses (**Figure 6**) show mainly the decrease of E′ defect component in the EPR spectra. Associated to this decrease, an increase in the relative proportion of OXY center relative to relative to BOHC defect is also detected.

The experiments testify that glass doping processes can influence the proportion between different defects produced during irradiation compared to the non-doped glass composition. This change has been usually correlated to the capacity of the doping ion to act as a trap for the holes and electrons produced during irradiation. Eu^{3+} ions are known to be good electron traps and one can observe on **Figures 1** and **5a** the strong decrease of the E′ proportion relatively to the non-doped ABS glasses. Sm^{3+} ion can also be reduced during the exposure to ionizing radiation but with a weaker efficiency than Eu^{3+} ions, this effect is also observed on the decrease of the relative proportion of E′ centers in the defect EPR spectra (**Figures 1** and **6**). By contrast, Ce^{3+} ions produced during irradiation or during the synthesis of Ce-doped glasses is a well-known hole traps [32, 33]. So these ions (Eu^{3+}, Sm^{3+}, and Ce^{3+}) can therefore compete with the hole trap defects production as represented for Eu-, Sm-, and Ce-doped ABS glasses. Relative proportion between hole and electron defects in the EPR spectra of irradiated RE-doped ABS glasses can be considered as a parameter for the estimation of the interaction of doping ions with the ionizing radiation.

However, other parameters can influence the nature of the different defects produced during irradiation and more specifically the speciation of the RE ions inside the host glasses. Indeed, some authors like Li and coworkers have studied the solubility and the environment of gadolinium in borosilicate glass compositions [34, 35]. They show that this RE ion is preferentially located in the vicinity of boron network former. According to literature, this RE ion is not therefore homogeneously distributed inside glass. This result influence strongly the EPR spectra of irradiated

Gd-doped ABS glasses (**Figure 1**) where the defect EPR spectrum is broad and not resolved due to dipole-dipole interaction between BOHC centers and paramagnetic Gd^{3+} ions. This effect could also explain the evolution of the EPR spectra of Eu-doped ABS glasses (**Figures 5a** and **5b**) as a function of doping ion content. This figure shows both decrease of the relative proportion of BOHC and E′ centers contributions in the defect EPR spectra. As Eu ions can be considered as an electron trap, the changes in relative proportion between OXY and BOHC defects shown in **Figure 6a** could show Eu speciation in the vicinity of boron network former. In order to understand the RE doping influence on defect production under ionizing radiation, Magic Angle Spinning Nuclear magnetic resonance spectroscopy (MAS NMR) of ^{11}B could be a way for studying as a function of RE nature in glasses, their possible influence on BOHC defect production.

The second influence of RE doping inside ABS glasses concerns the efficiency of defect production as a function of both nature of the doping ions and its content (**Figure 3**). The decrease of paramagnetic defects concentration can be due to the fact that the electron-hole pairs produced during ionizing radiation can support dynamical balance between the two different charge states of RE ions ($RE^{n+1} + (h°/e^-) => RE^n + h° => RE^{n+1}$ or $RE^n + (h°/e^-) => RE^{n+1} + e^- => RE^n$). Therefore, this process efficiency of defect production can be correlated to the stability of different oxidation states for different RE ions. Moreover, it is known that increase of the dopant content up to 1 mol.% in highly irradiated (3×10^9 Gy) Fe^{3+}- and Cr^{3+}-doped glasses may lead to the complete disappearance of the defect EPR spectrum [29, 30]. But it is necessary to take into consideration the Fe^{3+} (Cr^{3+}) ions dipole-dipole interaction effect on the defect EPR spectrum. In that case, the disappearance or decrease of defect EPR spectrum in the doped glasses as a function of doping ion content could also be associated with heterogeneous speciation of defects produced during exposure to ionizing radiation in the vicinity of doping ions.

For all RE doping ABS glasses considered in this work, saturation behavior of defect EPR spectra is analyzed as the function of the integrated dose. This result is explained by the strong efficiency of the dynamical trapping processes of electron-hole pairs on different redox states of RE doping ions with respect to the defect production efficiency under the effect of ionizing radiation in this ABS glass. In addition in **Figure 2a** and **b**, the defect content is decreasing in some cases at integrated doses higher than 5×10^8 Gy. Structural changes were observed for irradiated glass samples at doses around 10^9 Gy due to the effect of ionizing radiation and might be correlated to the alkaline migration [31]. The decrease of defect content at higher doses could therefore show an important role of precursor defect on the alkaline migration processes leading to glass structural changes.

5. Optical spectra: Effect of RE doping on defect band

The available structural information on defects in glass was derived mainly from the results of electron paramagnetic resonance (EPR) spectrometry. It should be noticed that this method is directly applicable only to the subclass of defects which are paramagnetic. A more formidable problem is the pre-existing intrinsic point defects, which are not of paramagnetic nature. Examples of intrinsic diamagnetic defects believed to occur in silica glass include neutral oxygen vacancies (≡Si-Si≡), two-coordinated silicone (O–Si–O–), and peroxy linkages (≡Si–O–O–Si≡) [5]. The most common extrinsic defects are associated with hydroxyl and chloride impurities [16]. It is obvious that the defect designation in multicomponent glasses is extremely complex. Thus, the combined information obtained from EPR- and optical (absorption and photoluminescence (PL)) spectra can give an additional data on the structure of the glasses and of the pre-existing/radiation-induced imperfections. Often some defects can provide large EPR signal but the induced optical extinction (transmission loss) is very low and vice versa.

A compendium of EPR/optical correlations was reported in the literature [36, 37] and pointed to the most likely origins of many defect-related optical absorption bands in the visible, ultraviolet, and vacuum-ultraviolet spectral regions. But the assignment of the bands is still controversial in some cases. In this section, some preliminary results on optical study of pristine and irradiated ABS glass doped with RE ions are presented.

Non-doped ABS glass has high ultraviolet transmission. No significant defect generation was detected (**Figures 8** and **9**, black line) in pristine glass. Only small transmission losses in the UV spectral region 230–240 nm were found (**Figure 9**) to be, it seems, connected with oxygen-deficient centers formed on the basis of silicon [15]. Oxygen-deficient centers (ODC, "oxygen vacancies") are the natural type of intrinsic defects in non-stoichiometric silicon dioxide [38]. By the existence of these types of defect in silica optical and luminescent properties are defined as described in [38, 39]. The dominating opinion has been still to consider vacancies of bridging oxygen atoms as the precursors of radiation E′ centers [38, 39]. Thus, ODCs play the key role in E′ center formation and their concentrations in glass. In irradiated ABS glass (BK7 or Duran type) the silicon and boron related electron centers (SiEC and BEC are considered to be responsible for this absorption) [37]. Also after irradiation with high dose (more than 10^9 Gy) some additional transmission losses can be observed in visible part of the spectra at ~360 and 600 nm (**Figures 9** and **10**, red line) caused by intrinsic radiation defect generation. In [37], these bands are attributed to the BOHC defects in ABS glass.

Photoluminescence (PL) was detected in both pristine and irradiated ABS glasses. As aforementioned in the spectra of pristine ABS glass the ODCs are displayed in the form of an absorption band at 230–240 nm. According to [15] ODSs emission is observed at 280 and 450 nm. Under excitation of forth harmonic

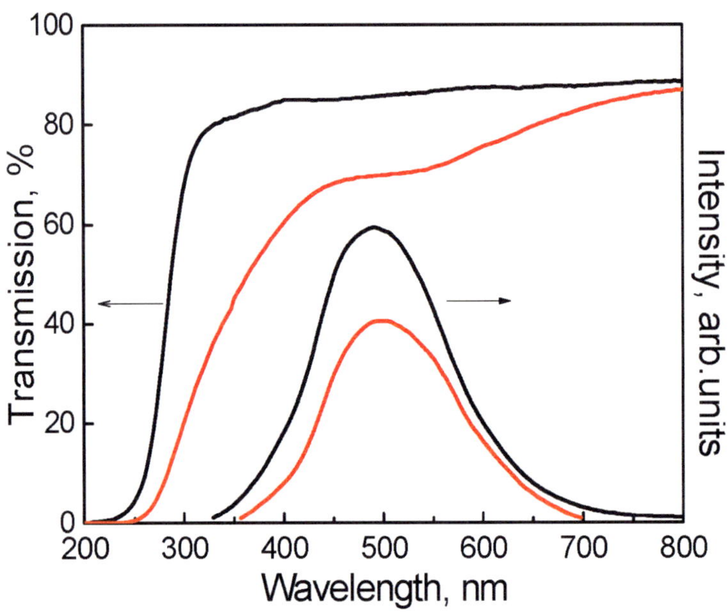

Figure 8.
Transmission and PL spectra of pristine and irradiated (10^9 Gy) non-doped ABS glasses.

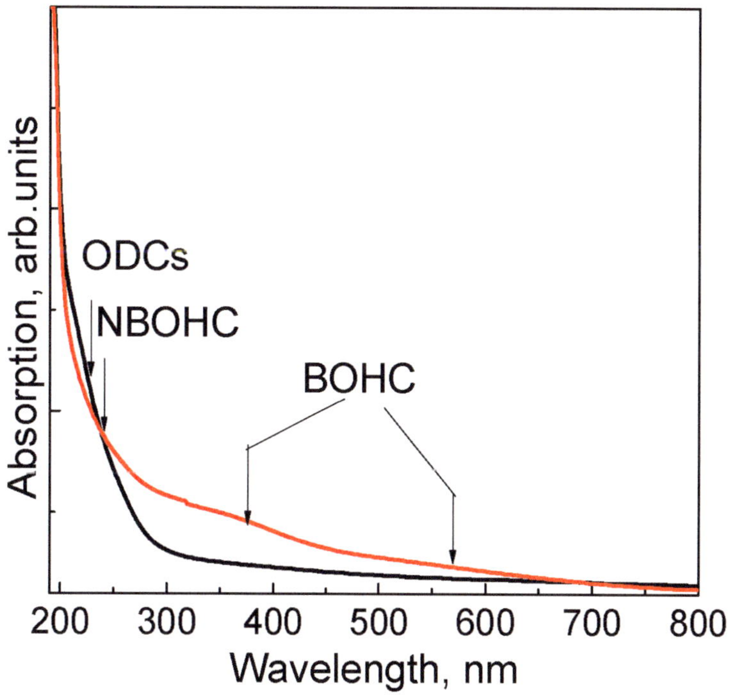

Figure 9.
Absorption spectra of pristine and irradiated (10^9 Gy) non-doped ABS glasses.

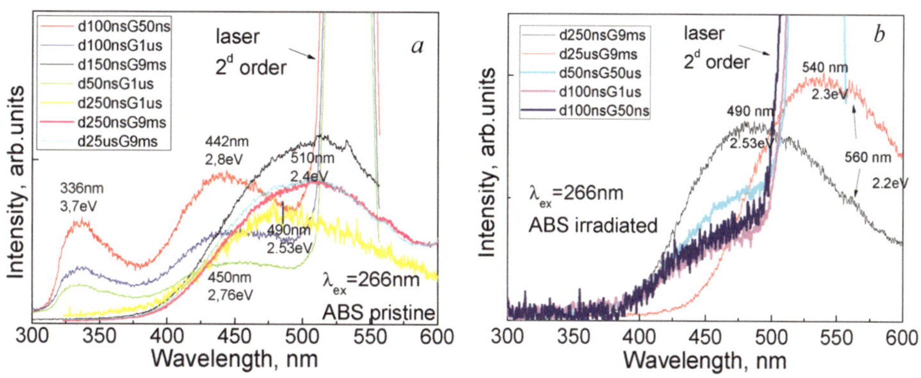

Figure 10.
Time-resolved luminescence spectra of pristine: (a) and irradiated (10^9 Gy); (b) non-doped ABS glasses measured at different gate width and time delay (λ_{exc} = 266 nm Nd:YAG laser).

of Nd:YAG laser (266 nm) we can see only the appearance of broad PL band at ~500 nm for both pristine (**Figure 8**, black line) and β-irradiated glass samples (**Figure 8**, red line). It should be noticed that the position, shape, width, and intensity of this band are different for these two glass samples (**Figure 8**).

Time-resolved luminescence measurements carried out with laser excitation (266 nm) revealed the variety of pre-existing point defects in ABS glass (**Figure 10a**) most of them are not identified and described in the literature on our opinion. For one exception: the band at 442 nm can be attributed to ODC which is consistent with data [15]. Especially taking into account that β-irradiation terminates this emission completely as well as two others at 336 and 510 nm. At the same time the new one (540 nm (5.3 eV); perhaps attributed to NBOHC [36]) is arising but emission band at 490 nm is still observed with higher intensity in irradiated ABS glass (**Figure 10b**). It should be marked here that the band at 336 nm can be also assigned to the ODCs since as it is indicated in [15] "Si ODCs in the silica glass network are an ensemble of defects such as 'oxygen vacancies', which differ in local structural environment, i.e., in the symmetry and strength of the local crystalline fields around the ODC." That is why the ODCs to be characterized by a rather wide variety of spectral characteristics.

The incorporation of RE ions into the ABS glass matrix affects its optical properties. The evolution of optical characteristics is discussed in frame of the non-bridging oxygen formation in the glass structure, as well as color centers as a function of the nature of the RE element. By increasing the number of Nd (Gd) ions in ABS glass, it is possible to observe a decrease in the number of non-bridging oxygen per silicon tetrahedron in the glass studied which is confirmed by the estimate of the optical band gap and the Raman spectroscopy data. The presence of two charge states of multivalent Eu and Ce ions having absorption in the UV region complicates the consideration of the effect of the processes on the observed change in the optical band gap energy. The effect of irradiation results in color centers content increase (observed as more intense brown coloration of the irradiated glass) in the following sequence, Nd < Gd < Sm = Eu < Ce. This evolution is reflected on the optical band gap narrowing [40]. **Figure 11** presents transmission and PL spectra of the highly Sm-doped (1 mol.%) pristine and irradiated ABS glass. Some transmission losses can be observed for pristine glass from **Figure 12**: firstly, due to the presence of Sm^{3+} ions in glass structure (sharp lines in the spectra) and secondly, due to the presence of ODCs (similar to the non-doped ABS glass, see **Figure 8**). The emission spectra consist of the well-known bands belonging to the Sm^{3+}/Sm^{2+} ions and the broad band around 500 nm, attributed to the ODCs: intensity of this band decreases significantly by β-irradiation (**Figure 11**).

The transmission spectra of Gd-doped ABS glass do not change significantly in comparison with non-doped glass except the fact that no prominent transmission losses are observed at 360 and 600 nm in irradiated glass (**Figure 12**). Emission of ODCs are located in the visible part of PL spectra at ~500 nm as it was seen before

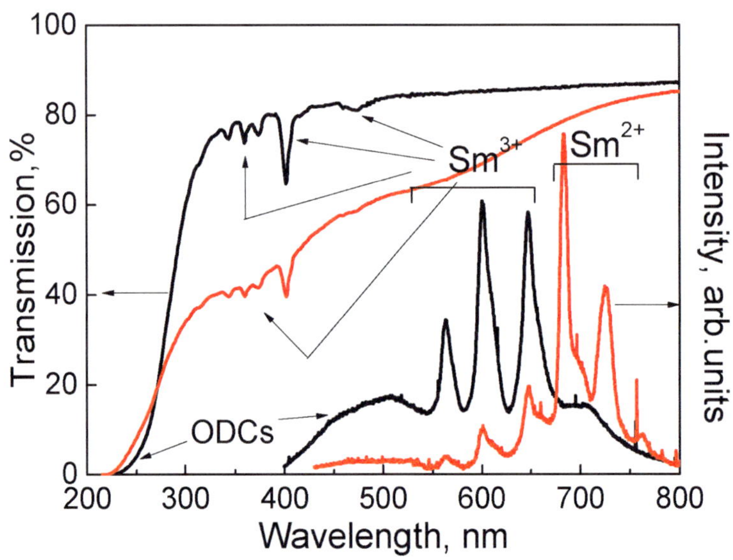

Figure 11.
Transmission and PL spectra of pristine and irradiated (10^9 Gy) Sm-doped ABS glasses.

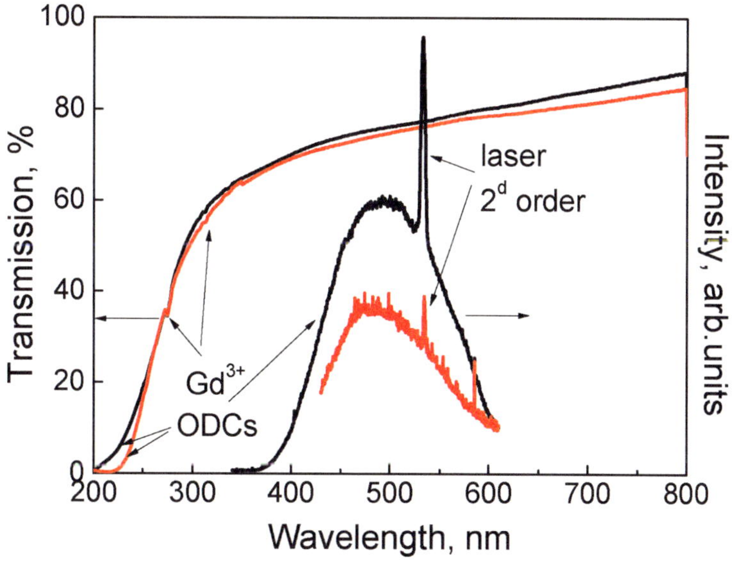

Figure 12.
Transmission and PL spectra of pristine and irradiated (10^9 Gy) Gd-doped ABS glasses.

for non-doped and Sm-doped ABS glass (**Figures 8** and **11**, respectively). It was shown [41] that this emission can also be excited by third harmonic of Nd:YAG laser (355 nm). Unfortunately, no information about ODCs optical characteristics could be obtained in Eu- or Ce-doped ABS glasses because of strong Eu^{2+}, $Ce^{3+}/^{4+}$ absorption in the UV studied spectral region [40]. It is clearly seen that the absorption and PL characteristics of ODCs in non-doped ABS glass, as well as in glass doped with Sm or Gd ions, are analogous to each other with the exception of unresolved PL bands structure in case of doping (**Figure 13**).

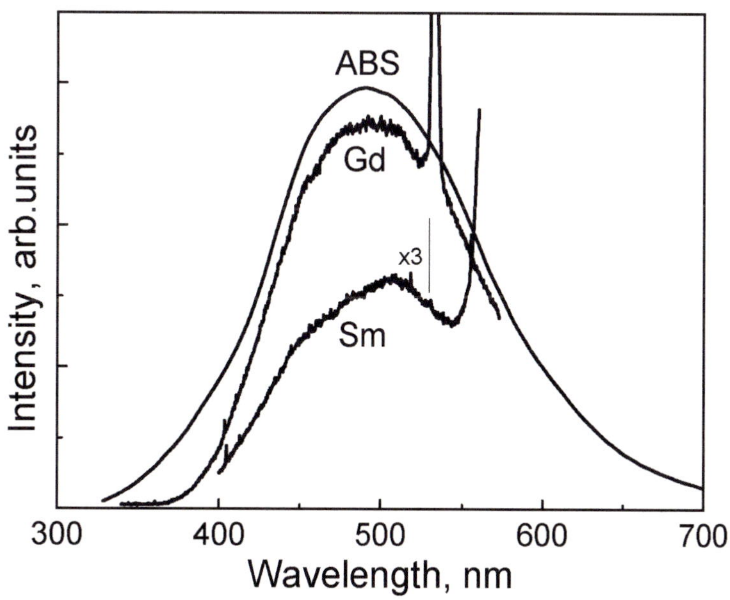

Figure 13.
Evolution of the defect emission band on RE dopant nature.

It was natural to suppose that we have similar types of defect. But for the moment it is not absolutely clear whether these ODCs have essentially different structure or whether they are the same vacancies with distorted environment. Moreover, the modified shape of the PL band can be consistent with the assumption about the effect of segregation of defects and impurities that can influence the variations of the spectral characteristics of the ODCs. In the previous section, the relative increase in the proportion of peroxy radicals (Si–O–O°: Oxy defects) with the Eu concentration as well of the Oxy and the HC_1 (Si–O Na^+) centers with Ce concentration (**Figures 4** and **5**) was reported. Probably it means that the structural model of the OXY and HC_1 centers tightly depends on the structural model of the ODC and it is still open for discussion. As to our point of view observed experimental facts can be most naturally explained by segregation of impurities and defects in the glass network. The effect of selective incorporation of a RE dopant into the glass due to the heterogeneous glass structure, leading to a RE concentration dependent dopant displacement as well as concentration dependent optical and physical-chemical glass properties which was firstly mentioned in the 1970 [42]. Then phase-separation model was suggested in order to explain structural evolution of RE-doped borosilicate glass [34, 35].

Analysis of the results presented allowed to draw a conclusion that due to RE speciation in the ABS glass structure and heterogeneous distribution between different environments RE doping affects strongly defect production (firstly, pre-existing defects). Additional studies on RE concentration dependent time-resolved luminescence ABS glass might be required.

6. Summary and outlook

The presented study has shown significant changes in defect creation processes under ionizing radiation in ABS glasses as a function of the nature and the content of different rare earth-doping ions (RE = Sm, Gd, Eu, Ce, and Nd). We observe first that doping processes influence significantly the nature of the different defect produced during ionizing radiation and more specifically the ratio between hole and electron defect centers observed by EPR spectroscopy. The specific role of RE doping ion acting as a hole or an electron trap could control the population of different defects produced during irradiation. The second result of doping is a strong decrease in defect production efficiency under the effect of ionizing radiation depending on both the RE doping content in the glass and on the relative stability of the RE different oxidation states. This result could be explained by dynamical reversible trapping of the electron-hole pairs produced during irradiation on the RE ions as well as by RE segregation and pre-existing defects speciation in ABS glass structure.

In order to understand the RE doping influence on defect production under the effect of ionizing radiation, Magic Angle Spinning Nuclear magnetic resonance spectroscopy (MAS NMR) of ^{11}B could be a way for studying as a function of RE nature in glasses and their possible influence on BOHC defect production.

Acknowledgements

I would like to thank Thierry Pouthier and Vincent Métayer for their contribution during external β-irradiation experiments. I do appreciate also to Dr. B. Boizot for stimulating my interest in this field and for pointing to my attention several interesting experimental problems as well as for many enlightening discussions.

Author details

Eugenia Malchukova
Ioffe Institute, Russian Academy of Sciences, St. Petersburg, Russia

*Address all correspondence to: e.malchukova@mail.ioffe.ru

References

[1] Pacchioni G, Skuja L, Griscom DL, editors. Defects in SiO_2 and Related Dielectrics: Science and Technology. Dordrecht: Kluwer Academic Publishers; 2000. DOI: 10.1007/978-94-010-0944-7. 615p

[2] Grillanda S, Singh V, Raghunathan V, Morichetti F, Melloni A, Kimerling L, et al. Gamma radiation effects on silicon photonic waveguides. Optics Letters. 2016;**41**(13):3053-3056. DOI: 10.1364/OL.41.003053

[3] Lisovskyy IP, Voitovych MV, Voitovych VV, Khacevich IM. Influence of radiation on the luminescence of silicon nanocrystals embedded into SiO_2 film. Journal of Nanomaterials. 2016;**2016**:9674741. DOI: 10.1155/2016/9674741

[4] Alessia A, Richard N, Martin-Samos L, De Michelea V, Giacomazzi L, Agnello S, et al. Overview of radiation induced point defects in silica-based optical fibers. Physical Review. 2019;**4**:100032(18). DOI: 10.1016/j.revip.2019.100032

[5] Griscom DL. Optical properties and structure of defects in silica glass. Journal of the Ceramic Society of Japan. 1991;**99**:923-942

[6] Du Q, Huang Y, Ogbuu O, Zhang W, Li J, Singh V, et al. Gamma radiation effects in amorphous silicon and silicon nitride photonic devices. Optics Letters. 2017;**42**(3):587-590. DOI: 10.1364/OL.42.000

[7] Skuja L, Hirano M, Hosono H, Kajihara K. Defects in oxide glasses. Physica Status Solidi C. 2005;**2**(1):15-24. DOI: 10.1002/pssc.200460102

[8] Imai H, Hirashima H. Intrinsic- and extrinsic-defect formation in silica glasses by radiation. Journal of Non-Crystalline Solids. 1994;**179**:202-213. DOI: 10.1016/0022-3093(94)90698-X

[9] Devine RAB, Arndt J. Correlated defect creation and dose-dependent radiation sensitivity in amorphous SiO_2. Physical Review B. 1989;**39**(8):5132-5138. DOI: 10.1103/physrevb.39.5132

[10] Devine RAB, Arndt J. Defect pair creation through ultraviolet radiation in dense, amorphous SiO_2. Physical Review B. 1990;**42**(4):2617-2620. DOI: 10.1103/physrevb.42.2617

[11] Weeks RA. The many varieties of E' centers: A review. Journal of Non-Crystalline Solids. 1994;**179**:1-9. DOI: 10.1016/0022-3093(94)90680-7

[12] Griscom DL, Friebele EJ. Fundamental defect centers in glass: 29Si hyperfine structure of the nonbridging oxygen hole center and the peroxy radical in a-SiO_2. Physical Review B. 1981;**2**(8):4896-4898. DOI: 10.1103/PhysRevB.24.4896

[13] Imai H, Arai K, Isoya J, Hosono H, Abe Y, Imagawa H. Generation of E' centers and oxygen hole centers in synthetic silica glasses by gamma irradiation. Physical Review B. 1993;**48**(2):3116-3123. DOI: 10.1103/physrevb.48.3116

[14] Zhang L, Mashkov VA, Leisure RG. Multiple interconversions of the E' and oxygen-hole defect centers in high-purity amorphous silica during anneal-interrupted x irradiation. Physical Review Letters. 1995;**74**(9):1605-1608. DOI: 10.1103/PhysRevLett.74.1605

[15] Amossov AV, Rybaltovsky AO. Radiation color center formation in silica glasses: A review of photo- and thermo-chemical aspects of the problem. Journal of Non-Crystalline Solids. 1994;**179**:226-234. DOI: 10.1016/0022-3093(94)90700-5

[16] Skuja L, Kajihara K, Hirano M, Hosono H. Hydrogen-related radiation

defects in SiO_2-based glasses. Nuclear Instruments and Methods in Physics Research B. 2008;**266**(12-13):2971-2975. DOI: 10.1016/j.nimb.2008.03.150

[17] Elsts E, Rogulis U, Bulindzs K, Smits K, Zolotarjovs A, Trinkler L, et al. Studies of radiation defects in cerium, europium and terbium activated oxyfluoride glasses and glass ceramics. Optical Materials. 2015;**41**:90-93. DOI: 10.1016/j.optmat.2014.10.042

[18] Uklein AV, Popov AS, Lisnyak VV, Zaderko AN, Linnik RP, Boldyrieva OY, et al. Probing of the oxygen-related defects response in Nd:phosphate glass within self-action of the laser radiation technique. Journal of Non-Crystalline Solids. 2018;**498**:244-251. DOI: 10.1016/j.jnoncrysol.2018.06.024

[19] Mashkov VA, Austin WR, Zhang L, Leisure RG. Fundamental role of creation and activation in radiation-induced defect production in high-purity amorphous SiO_2. Physical Review Letters. 1996;**76**(6):2926-2929. DOI: 10.1103/PhysRevLett.76.2926

[20] Griscom DL. E.S.R. studies of radiation damage and structure in oxide glasses not containing transition group ions: A contemporary overview with illustrations from the alkali borate system. Journal of Non-Crystalline Solids. 1973;**13**:251-285. DOI: 10.1016/0022-3093(74)90095-7

[21] Kordas G. On the structure of the BOHC in the borosilicate and boro-phosphosilicate glasses. Journal of Non-Crystalline Solids. 2005;**351**:2348-2360. DOI: 10.1016/j.jnoncrysol.2005.06.010

[22] Dutt DA, Higby PL, Griscom DL. An electron spin resonance study of X-irradiated calcium aluminosilicate glasses. Journal of Non-Crystalline Solids. 1991;**130**:41-521. DOI: 10.1016/0022-3093(91)90154-X

[23] Kordas G, Camara B, Oel HJ. Electron spin resonance studies of radiation damage in silicate glasses. Journal of Non-Crystalline Solids. 1982;**50**:79-95. DOI: 10.1016/0022-3093(82)90202-2

[24] Griscom DL. Electron spin resonance studies of trapped hole centers in irradiated alkali silicate glasses: A critical comment on current models for HC_1 and HC_2. Journal of Non-Crystalline Solids. 1984;**64**:229-247. DOI: 10.1016/0022-3093(84)90220-5

[25] Boizot B, Petite G, Ghaleb D, Calas G. Radiation induced paramagnetic centers in nuclear glasses by EPR spectroscopy. Nuclear Instruments and Methods in Physics Research B. 1998;**141**:580-584. DOI: 10.1016/S0168-583X(98)00102-5

[26] Malchukova E, Boizot B, Petite G, Ghaleb D. Optical properties and valence state of Sm ions in aluminoborosilicate glass under β-irradiation. Journal of Non-Crystalline Solids. 2007;**353**: 2397-2402. DOI: 10.1016/j.jnoncrysol.2007.04.003

[27] Malchukova E, Boizot B, Ghaleb D, Petite G. β-Irradiation effects in Gd-doped borosilicate glasses studied by EPR and Raman spectroscopies. Journal of Non-Crystalline Solids. 2006;**352**(4):297-303. DOI: 10.1016/j.jnoncrysol.2005.11.003

[28] Ollier N, Planchais R, Boizot B. EPR study of Yb-doped irradiated glasses. Nuclear Instruments and Methods in Physics Research B. 2008;**266**(12-13): 2854-2858. DOI: 10.1016/j.nimb.2008.03.129

[29] Olivier FY, Boizot B, Ghaleb D, Petite G. Raman and EPR studies of β-irradiated oxide glasses: The effect of Iron concentration. Journal of Non-Crystalline Solids. 2005;**351** (12-13):1061-1066. DOI: 10.1016/j.jnoncrysol.2005.01.018

[30] Boizot B, Olivier FY, Petite G, Ghaleb D. Blocking of alkaline migration under ionizing irradiation in Cr-doped oxide glasses. Nuclear Instruments and Methods in Physics Research B. 2008;**266**(12-13):2966-2970. DOI: 10.1016/j.nimb.2008.03.149

[31] Boizot B, Ollier N, Olivier F, Petite G, Ghaleb D, Malchukova E. Irradiation effects in simplified nuclear waste glasses. Nuclear Instruments and Methods in Physics Research B. 2005;**240**(1-2):146-151. DOI: 10.1016/j.nimb.2005.06.105

[32] Qiu J, Sugimoto N, Iwabuchi Y, Hirao K. Photostimulated luminescence in Ce^{3+}-doped silicate glasses. Journal of Non-Crystalline Solids. 1997;**209**:200-203. DOI: 10.1016/S0022-3093(96)00644-8

[33] Vedda A, Nikl M, Fasoli M, Mihokova E, Pejchal J, Dusek M, et al. Thermally stimulated tunneling in rare-earth-doped oxyorthosilicates. Physical Review B. 2008;**78**:195123(8). DOI: 10.1103/PhysRevB.78.195123

[34] Li L, Li H, Quian M, Strachan DM. Gadolinium solubility in peralkaline borosilicate glasses. Journal of Non-Crystalline Solids. 2001;**283**:237-245. DOI: 10.1016/S0022-3093(01)00480-X

[35] Li H, Su Y, Li L, Strachan DM. Raman spectroscopic study of gadolinium(III) in sodium-aluminoborosilicate glasses. Journal of Non-Crystalline Solids. 2001;**292**:167-176. DOI: 10.1016/S0022-3093(01)00878-X

[36] Skuja L. Optically active oxygen-deficiency-related centers in amorphous silicon dioxide. Journal of Non-Crystalline Solids. 1998;**239**(1-3):16-48. DOI: 10.1016/S0022-3093(98)00720-0

[37] Ehrt D, Ebeling P. Radiation defects in borosilicate glasses. European Journal of Glass Science and Technology. Part A: Glass Technology. 2003;**44**(2):46-49

[38] Weeks RA. Paramagnetic resonance of lattice defects in irradiated quartz. Journal of Applied Physics. 1956;**27**(11): 1376-1381. DOI: 10.1063/1.1722267

[39] Hosono H, Abe Y, Kinser DL, Weeks RA, Muta K, Kawazoe H. Nature and origin of the 5-eV band in SiO_2:GeO_2 glasses. Physical Review B. 1992;**46**:11445-11451. DOI: 10.1103/physrevb.46.11445

[40] Malchukova EV, Nepomnyashchikh AI, Boizot B, Terukov EI. Radiation effects and optical properties of aluminoborosilicate glass doped with RE ions. Glass Physics and Chemistry. 2018;**44**(4):356-363. DOI: 10.1134/S1087659618040090

[41] Mal'chukova EV, Nepomnyashchikh AI, Boizot B, Shamirzaev TS, Petite G. Luminescence of aluminoborosilicate glasses doped with Gd^{3+} ions. Physics of the Solid State. 2010;**52**(9):1919-1924. DOI: 10.1134/S1063783410090222

[42] Dmitryuk AV, Karapetyan GO, Maksimov LV. Activator segregation and its spectroscopic signatures. Zhurnal Prikladnoi Spektroskopii. 1975;**22**(1):153-182